# Nanotechnology in the Life Sciences

**Series Editor**
Ram Prasad, Department of Botany
Mahatma Gandhi Central University
Motihari, Bihar, India

Nano and biotechnology are two of the 21st century's most promising technologies. Nanotechnology is demarcated as the design, development, and application of materials and devices whose least functional make up is on a nanometer scale (1 to 100 nm). Meanwhile, biotechnology deals with metabolic and other physiological developments of biological subjects including microorganisms. These microbial processes have opened up new opportunities to explore novel applications, for example, the biosynthesis of metal nanomaterials, with the implication that these two technologies (i.e., thus nanobiotechnology) can play a vital role in developing and executing many valuable tools in the study of life. Nanotechnology is very diverse, ranging from extensions of conventional device physics to completely new approaches based upon molecular self-assembly, from developing new materials with dimensions on the nanoscale, to investigating whether we can directly control matters on/in the atomic scale level. This idea entails its application to diverse fields of science such as plant biology, organic chemistry, agriculture, the food industry, and more.

Nanobiotechnology offers a wide range of uses in medicine, agriculture, and the environment. Many diseases that do not have cures today may be cured by nanotechnology in the future. Use of nanotechnology in medical therapeutics needs adequate evaluation of its risk and safety factors. Scientists who are against the use of nanotechnology also agree that advancement in nanotechnology should continue because this field promises great benefits, but testing should be carried out to ensure its safety in people. It is possible that nanomedicine in the future will play a crucial role in the treatment of human and plant diseases, and also in the enhancement of normal human physiology and plant systems, respectively. If everything proceeds as expected, nanobiotechnology will, one day, become an inevitable part of our everyday life and will help save many lives.

Vishnu Kirthi Arivarasan
Karthik Loganathan • Iddya Karunasagar
Editors

# Nanotechnological Approaches to the Advancement of Innovations in Aquaculture

Springer

*Editors*
Vishnu Kirthi Arivarasan
Department of Microbiology
School of Bioengineering and Biosciences
Lovely Professional University
Phagwara, Punjab, India

Karthik Loganathan
Department of Synthetic Biology
Salem Microbes
Salem, Tamil Nadu, India

Iddya Karunasagar
Center for Science Education & Research
Nitte University
Kotekar, Karnataka, India

ISSN 2523-8027 ISSN 2523-8035 (electronic)
Nanotechnology in the Life Sciences
ISBN 978-3-031-15518-5 ISBN 978-3-031-15519-2 (eBook)
https://doi.org/10.1007/978-3-031-15519-2



This Springer imprint is published by the registered company Springer Nature Switzerland AG
The registered company address is: Gewerbestrasse 11, 6330 Cham, Switzerland

# Preface

Nanotechnology is a cutting-edge science and technology with numerous scientific and technological applications. Rapid breakthroughs in nanosciences and nanotechnologies have opened up new possibilities for many industrial and consumer sectors, including aquaculture and agriculture and associated fields, which have recently been considered a hotspot of a new industrial revolution.

Aquaculture, or the farming of aquatic organisms in inland and coastal settings, is the world's fastest expanding food-generating sector, thanks to technological advancements and diversification. It is widely perceived that the high-quality proteins found in fish are superior to that found in meat and fowl. Annually, the aquaculture industry generates substantial advances in the creation of technology to boost food production.

Nanotechnology has played an important role in the development of an economic and sustainable pathway for different aspects of aquaculture for benefit of the human race. It has led to the development of new tools for aquaculture, fish biotechnology, fish genetics, fish reproduction, and aquatic health, among other things. The fisheries and aquaculture industries can be changed by integrating nanotechnology with new tools such as rapid disease diagnosis, improving fish's ability to absorb medications such as hormones, vaccinations, and nutrients, and so on.

In this book, we have tried to put together a detailed account of the different roles nanotechnology plays in the fast-developing and progressive path toward aquaculture growth. The different aspects of nanotechnology in aquaculture have been amalgamated into a single book. This book will be beneficial for young researchers, postgraduates, and of course the general populace to know in detail about the topic of nanotechnology and aquaculture.

Phagwara, Punjab, India — Vishnu Kirthi Arivarasan
Salem, Tamil Nadu, India — Karthik Loganathan
Kotekar, Karnataka, India — Iddya Karunasagar

# Contents

# Chapter 1
# Nanotechnologies in Aquatic Disease Diagnosis and Drug Delivery

**V. Baskaran**

## Abbreviations

| | |
|---|---|
| ALP | Alkaline phosphatase |
| CNTs | Carbon nanotubes |
| DDS | Drug delivery system |
| DNA | Deoxyribonucleic acid |
| EMA | European Medicine Agency |
| FAT | Fluorescence antibody test |
| GCRC | Grass carp reovirus |
| GO | Graphene oxide |
| HK | Head kidney |
| IFAT | Indirect fluorescence antibody test |
| IHC | Immunohistochemistry |
| IHNV | Infectious hematopoietic necrosis virus |
| IM | Intramuscular |
| IP | Intraperitoneal |
| IPNV | Infectious pancreatic necrosis virus |
| ISAV | Infectious salmon anaemia virus |
| KHV | Koi herpes virus |
| LCDV | Lymphocystis disease virus |
| LPS | Lipopolysaccharides |
| NP's | Nanoparticles |
| PCR | Polymerase chain reaction |
| PLA | Poly lactic acid |
| PLGA | Poly (lactic-co-glycolic acid) |
| RAPD | Random amplification of polymorphic DNA |
| rcb-PCR | Reverse cross blot PCR |
| RNA | Ribonucleic acid |
| ROS | Reactive oxygen species |

V. Baskaran (✉)
Salem Microbes Private Limited, Salem, Tamil Nadu, India



V. K. Arivarasan et al. (eds.), *Nanotechnological Approaches to the Advancement of Innovations in Aquaculture*, Nanotechnology in the Life Sciences, https://doi.org/10.1007/978-3-031-15519-2_1

| | |
|---|---|
| RT-PCR | Reverse transcriptase polymerase chain reaction |
| SLN | Solid lipid nanoparticles |
| TLRs | Toll-like receptors |
| VHSV | Viral haemorrhagic septicaemia virus |
| WSSV | White spot syndrome virus |

## 1.1 Introduction

Aquaculture represents the fastest-growing domain with prolific growth potential. It serves as the source of easily digestible protein, healthy fat, and a cache of various essential micronutrients for billions across the globe. Aquaculture provides a potential platform for rural employment and livelihood development, possesses high export earning potential, and can support national GDP through income generation. Recent years have seen development in the production of farmed fin and shellfish that has increased rapidly due to enhanced, intensive aquaculture practices (FAO 2018) and demand. Due to the increase in intensive farming systems, the need for wild seed stocks as well as fish supplies has declined (Naylor et al. 2006). A report from the Food and Agriculture Organization shows that the average annual consumption of fish increased by 3.2% globally from 1961 to 2016. In 2015, of the total animal protein consumed globally, fish food dominated at 17% and supplied almost 20% of the average consumption of animal protein for 3.2 billion populations (FAO 2018). However, environmental factors like excessive accumulation of nutrients result in the eutrophication of ponds due to intensive aquaculture practices, which are being looked at as a block in aquaculture growth dynamics (Herbeck et al. 2013). Along with eutrophication, climate change also poses a serious threat to aquaculture due to the temperature rise, increased methane and carbon dioxide emissions, etc. (FAO 2018). An increase in water temperature will have an impact on plankton communities, the survival of larvae and juveniles, and the reproductive ability of fishes. It has been also observed that contaminated water is being introduced into reservoirs and fish culture systems (Patel et al. 2019). Moreover, detecting pollution and toxic levels in water is a tedious and time-consuming process (Altenburger et al. 2019). In extensive aquaculture practices, nutritional deficiency is common among juvenile and broodfish populations, and health management is also another challenging problem. Depleting environmental quality and the impact of climate change have led to a surge in infectious diseases in aquaculture. Diseases like fin and gill rot and epizootic ulcerative syndrome (EUS) are still prevalent with no possible solutions and become incurable with antibiotics.

A white spot disease outbreak has inhibited the otherwise profitable *Penaeus monodon* culture system (Zhang et al. 2016). It has been estimated that the global shrimp industry has faced a loss of US$10 billion from white spot disease syndrome and the infectious myonecrotic virus alone since 1990. Meanwhile, the $15 billion global ornamental fish business also faced a problem with antibiotic resistance, which created uncertainty in its therapeutic practices. Conventional drug delivery models have low efficacy and bioavailability in aquatic ecosystems. Besides these, traditional disease detection methods are also not feasible in the long run.

At this point, there is a need for some innovative technological interventions to get rid of these problems in aquaculture. Nanotechnology has emerged as a promising solution, leading to new possibilities by developing and applying materials at nanoscale dimensions (1–100 nm) with unique properties, paving the way for novel therapeutic as well as diagnostic applications (Matteucci et al. 2018; Mohamed et al. 2015). The application of nanotechnology has become a part of human life with multiple value-added products (Stern et al. 2016; Vinay et al. 2018). With recent technologies, nanoparticles are being prepared by different physical and chemical methods. The main commercial nanoparticles include silver, gold, oxides of zinc, iron, calcium, titanium, and manganese, as well as carbon nanotubes (CNTs), mesoporous silica nanoparticles, quantum dots (Khan et al. 2019; Stone et al. 2010), etc. With complex three-dimension technology, 'Second-generation' nanomaterials are now being synthesized with highly functional surfaces (Sarkar et al. 2022; Jeevanandam et al. 2018; Ju-Nam and Lead 2008). Even the impact of toxic pollutants on aquatic species can be intervened by using nanotechnology. Besides conventional toxicological assays, scientists are also using cellular cytotoxicity, apoptosis as well as bioinformatics-based interactive tools to narrate nanotoxicity and its remediation by applying green nanoparticles in bacteria and fish models (Kumari et al. 2017; Husain et al. 2021; Verma et al. 2017, 2021). Hence, the efficient application of nanotechnology can be addressed for maximizing aquaculture productivity and a schematic diagram is presented in this regard as Fig. 1.1.

## 1.2 Fish Disease and Immune System

The disease is a primary cause of fish mortality, especially when fishes are in the larval stage. There are two types of fish diseases according to their infectious pattern, viz. pathogenic diseases and non-pathogenic diseases (Arfat et al. 2014; Vevers and Jha 2008). Different infectious diseases are the main cause of death that might be due to bacterial, viral, or parasitic infection, and sometimes pathogens cause high levels of toxicity. It is mainly related to poor water quality management, inadequate feed supply, etc. A high-nutrition diet will have a positive effect on fish health. In this way, nanoparticles serve as a main source of transportation as well as supplementation of materials across the membrane. Non-infectious diseases are mainly caused by gas bubble diseases due to extensive aeration, nutritional diseases due to deficiency of nutrients such as vitamins and minerals, disorders caused by pollutants in the form of agricultural and industrial wastes, and neo-plastic and genetic anomalies resulting in abnormal growth which makes organs lose their structure and function (Rajkumar et al. 2016; Gao et al. 2017; Vinay et al. 2018; Bacchetta et al. 2017; Verma et al. 2020).

The fish immune system plays an important role in the protection against pathogens. Under the infectious stage, the immune system can detect pathogens

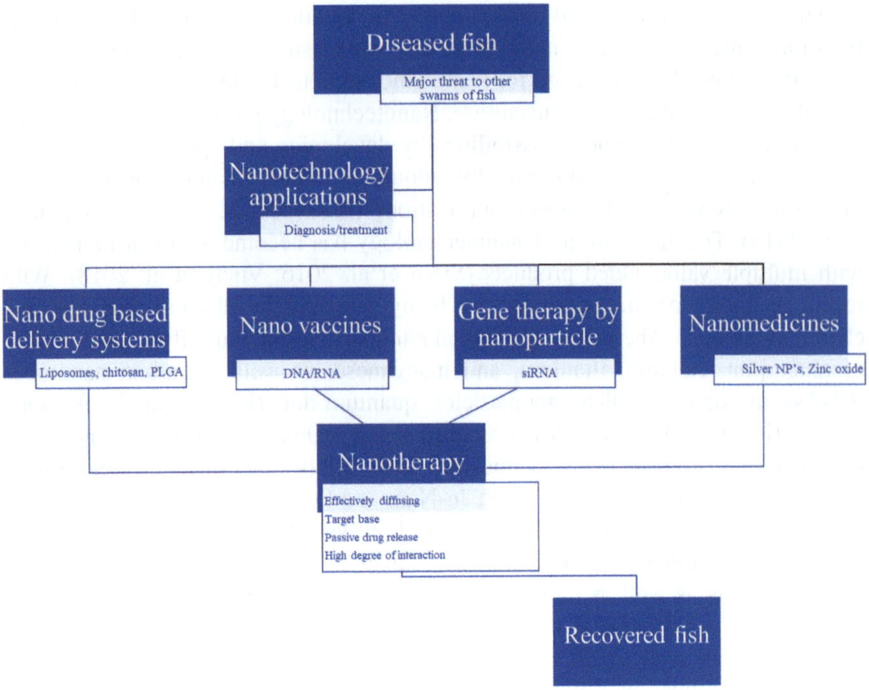


**Fig. 1.1** Application of nanotechnology in aquaculture

(Firdaus-Nawi and Saad 2016). The immune system of fish is composed of two main components: innate and adaptive immunities (Firdaus-Nawi and Saad 2016). Innate immunity is nonspecific and serves as the primary defence against pathogen invasion, while adaptive immunity is much more specific to a particular pathogen. Innate immunity is composed of nonspecific cellular and humoral elements. The nonspecific cellular portion consists of toll-like receptors (TLRs), macrophages, neutrophils, eosinophils, and nonspecific cytotoxic cells, while the nonspecific humoral portion includes lysozyme, complement cells, interferons, C-reactive proteins, transferrins, and lectins, where they work together to prevent pathogen invasion at the primary stage of infection (Firdaus-Nawi and Saad 2016). On the other hand, the adaptive immune system comprises highly specific systemic cells and mechanisms that are divided into two major components: humoral and cellular (Firdaus-Nawi and Saad 2016). Three forms of antibodies, IgM, IgD, and IgT, are essential components of humoral immunity which operate on invading extracellular diseases. The cytotoxic T-lymphocyte cells are a significant component of cellular immunity that often destroys invading bacterial, viral, or parasite-infected cells and intracellular cells (Firdaus-Nawi and Saad 2016). Both innate and adaptive immune systems work together efficiently to protect the body from any diseases, whereas innate immunity responds to invading pathogens by identifying line-encoded molecules of the germ.

## 1.3 Key Factors to Improve Disease Diagnosis and Vaccine Development

The major underlying factors for improving disease diagnosis and vaccine development are the continued significant losses to the industry caused by pathogens. Bacterial diseases cause substantial economic inflation in the aquaculture industry, and although antibiotics and chemotherapeutics are extensively used to control disease outbreaks, there is increasing concern about their use because of drug residues in food, the development of antimicrobial drug resistance, and their detrimental effect on aquatic microbial ecosystems and populations (Thompson and Adams 2004).

## 1.4 Limitations of Current Diagnostic Methods

Most of the current techniques for the detection of pathogens and the diagnosis of diseases are reliable. On the other hand, it is difficult to identify certain pathogens, and some of the methods developed may be too complicated to apply or infer. Conventional pathogen isolation and characterization techniques, combined with histopathology, remain the methods of choice for the diagnosis of many diseases. However, these traditional methods are costly, labour-intensive, slow, and time-consuming, and might not always lead to a definitive diagnosis being made. The rapid progress made in biotechnology since the 1990s has enabled the development and improvement of a wide range of immunodiagnostic and molecular techniques (Cunningham 2004; Adams and Thompson 2006, 2008), and reagents and commercial kits have become more widely available. These rapid methods both complement and enhance the traditional methods of disease diagnosis.

## 1.5 Advances in Methods of Disease Diagnosis

Disease diagnosis is currently made using a variety of methods, as reviewed by Adams and Thompson (2008). Traditional bacteriological methods, where the pathogen is isolated and identified biochemically (e.g. using API strips), and observation of histological sections from diseased fish are widely used. Rapid methods that specifically identify the pathogen using antibodies (immunodiagnostics) or by amplifying specific sequences of DNA or RNA using polymerase chain reaction (PCR) (i.e. molecular diagnostics) are also being used in many laboratories. In some cases, recent advances in molecular diagnostics have completely replaced other methodologies.

Immunodiagnostic methods such as immunohistochemistry (IHC), the fluorescence antibody test (FAT), and the indirect fluorescence antibody test (IFAT) enable rapid and specific detection of pathogens in tissue samples without the need to first isolate the pathogen. The use of molecular technologies for the detection of fish bacterial pathogens is rapidly increasing, and a vast array of methods has already been developed (Karunasagar et al. 1997; Cunningham 2004; Adams and Thompson 2006, 2008; Wilson and Carson 2003). Molecular methods generally have the highest sensitivity and are particularly useful for detecting microorganisms that are present in low copy numbers or those that are unculturable. The PCR is the best-known method, although there are many different kits, including nested PCR, random amplification of polymorphic DNA (RAPD), reverse transcriptase-PCR (RT-PCR), reverse cross blot PCR (rcb-PCR), and RT-PCR enzyme hybridization assay (Puttinaowarat et al. 2000; Wilson and Carson 2003; Cunningham 2004).

## 1.6 Nanotechnology Tool for Enhancing Aquaculture Operations and Fishing System

For any aquaculture practice, the pond is the core unit, and its professional management is the primary criterion for developing a successful aquaculture system. To design a perfect pond structure, multiple construction materials, as well as fabrication items, are required. Moreover, water supply and management of quality water are of primary importance. Sometimes, natural water bodies like the extensions of reservoirs and lakes, commonly called creeks, can be used for cage or open culture systems. The use of nanoparticles like silver, zinc oxide, copper oxide, titanium oxide, etc. can be mixed and coated with fabrication materials to increase the shelf life of the pond. Carbon nanotubes are approximately a hundred times stronger than steel and could be employed as an additive in boat and gear making or in designing cages and nets. ULVAC Inc., a corporate leader in vacuum equipment manufacturing, has designed a fishing lure coated with several hundred-nanometre-thick polyimide films that have a natural shining effect. Using this lure, better fishing practices can be achieved (Handy 2012).

## 1.7 Nanotechnology in Fish Disease Control

Disease occurrence is one of the major perils in intensive aquaculture practices, (Toranzo et al. 2005), and management of pond health is also important for the prevention and curing of fish species from pathogenic infestation. Nanotechnology can contribute significantly through novel methods as well as restructuring conventional technology (Kaul et al. 2018).

## 1.8 Nanomaterials as a Diagnostic Tool

In the modern era, aquaculture scientists are looking for an easy and reliable method for early detection and routine diagnostics of important fish diseases. An antibody-based, highly sensitive immunodiagnostics protocol incorporating nanoscale gold with alkaline phosphatase (ALP) conjugated secondary antibody was developed to measure titre against white spot syndrome virus (WSSV) in shrimp (Thiruppathiraja et al. 2011). Moreover, it has been reported that immuno-targeted gold nanoparticles are being conjugated with antibodies targeted to a particular biomolecule of interest, such as immunoglobulin G-capped gold nanoparticles, that bind specifically to antibodies produced against *Staphylococcus pyogenes* and *S. aureus*, among others (Roy et al. 2012).

Similarly, nanosensors also serve as an effective and easy method to identify pathogens. Nano biosensor systems are currently being developed for detecting very low concentrations not only of parasites, bacteria, and viruses but also of polluting elements in the water (Chen et al. 2016). Different nanosensors are being used to detect important aquaculture viruses like aquabirnavirus, salmonid alphavirus, and betanodavirus (Crane and Hyatt 2011). Currently, graphene oxide (GO) is used for designing electrochemical biosensors due to its unique chemical constitution and biocompatibility. A simple, sensitive, real-time, and rapid detection method of white spot syndrome virus (WSSV) in shrimp samples was demonstrated by applying a GO-based electrochemical immunosensor. The detection limit was found to be very low, at $1.36 \times 10^{-3}$ copies $\mu L^{-1}$. This is a unique and alternative qualitative and quantitative method, unlike PCR amplification-based detection techniques (Natarajan et al. 2017). This is particularly important in outbreaks at commercial aquaculture systems since conventional methods take too much time to identify the etiological agent causes, delaying the treatment to control the pathogen and creating an important economic impact. In this regard, nanotechnology holds the key to overcoming the potential threat through early detection and eradication of pathogens. Currently, nanosensors can detect a wide range of pathogens, be it bacteria, viruses, or parasites. For instance, electrical nanosensors are feasible to detect single virus particles (Patolsky et al. 2004). Nanobiosensors are also used for fishpond cleaning and stock inspections, such as those based on carbon nanotubes, which are highly sensitive for the detection of traces of pathogens like viruses, parasites, and bacteria, as well as heavy metals, in both food and water (Husain et al. 2021).

## 1.9 Nanomaterials as Nanomedicine

Nanomedicine, another rapidly growing area in nanotechnology, has a wide opportunity to use the intrinsic properties of different forms of nanoparticles for improving fish health. The antimicrobial and prophylactic actions of nanomaterials like nanosilver and zinc oxide nanoparticles are already exploited to reduce the pathogenic load in the aquaculture system (Siddiqi et al. 2018). This unique phenomenon is nonspecific, universal, and widely applicable. Antibacterial properties of nanoparticles such as titanium dioxide and copper oxide are being investigated as a potential nanomedicine for fish. Graphene appears as a commercially attractive, cheap, and renewable nanomaterial when, in its oxidized form, it is easy to process and dispersible in water (Brisebois and Siaj 2019). Graphene oxide (GO) exhibited an inhibitory effect against important aquatic pathogens like *S. aureus*, *P. aeruginosa*, *E. coli*, and *Vibrio harveyi*, both individually and in combination with metal/metal oxide and polymeric nanoparticles. Graphene oxide causes mechanical enfolding followed by cell membrane damage and lysis upon interacting with pathogens (Kumar et al. 2019).

## 1.10 Phyto-Nanocomposites as Nanomedicines

Different herbal and phytoextracts are currently being used as potential drugs in treating fish diseases. Different nanoparticles are being synthesized using medicinal plant and herbal extracts at optimized hydrodynamic conditions, and a composite of the phyto-nanoformulation is delivered as drugs with synergistic effects on improving fish health. Mahanty et al. (2013a, b) described some methods of phytoextract-synthesized nanosilver composite showing effective inhibition against the important fish pathogen, *Aeromonas hydrophila*.

## 1.11 Nanomaterials as Drug Delivery Systems (DDS) and Therapeutic Tools

Nano-delivery of drugs is attributed to novel properties like the sustained release, regulation, and control of size, shape, dispersity, and surface charge of targeted materials; location-specific, multi-route delivery processes; and regulated degradability of nanocarriers (Patra et al. 2018).

## 1.12 Nanovaccine Delivery

Nanotechnology, in conjunction with biotechnology, has made tremendous progress in the biomedicine field (Zhao et al. 2014) and has expanded its portfolio in the field of vaccinology, thereby giving rise to a new scientific field called nanovaccinology (Mamo and Poland 2012; Zhao et al. 2014). Vaccination plays a pivotal role in large-scale, commercial pisciculture and has been a major factor in the success of salmon and trout cultivation (FAO 2018) and partially in Indian major carp farming.

In the last 20 years, fish vaccinations have become a highly recognized, cost-effective method for preventing certain pathogenic infections in aquaculture (Assefa and Abunna 2018; Vinitnantharat et al. 1999). Most of the vaccines produced are kept and preserved in liquid form at low temperatures and are usually injected through blood networks due to their short life span. These bottlenecks have reduced the chances of vaccines being widely used, particularly in certain finfish and shellfish. 'Nanovaccine' is an emerging mass vaccination method in aquaculture (Assefa and Abunna 2018). The USDA demonstrated ultrasound-driven tools for mass vaccination of fish. Short strands of DNA-loaded nanocapsules are placed in the fish pond, where they adsorb onto the surface of fishes. Then ultrasound is triggered to rupture the capsules for releasing the DNA eliciting an immunological response. This technology has been examined on rainbow trout and commercialized by 'Clear Springs Foods', USA (Mongillo 2007). Similarly, oral delivery of vaccines and targeted release of active ingredients for vaccination will reduce the cost of fish farming (Aravena et al. 2013).

DNA nanovaccines, comprising short strands of DNA within nanocapsules, have been used in the aquaculture industry to induce an immune response in fishes. Iron nanoparticles have been shown to accelerate fish development when combined with drug delivery via programmed release (Hussain et al. 2021). For instance, Bhattacharyya et al. (2015) successfully showed nanoencapsulated vaccines against the bacterium *Listonella anguillarum* in Asian carp and also in rainbow trout (*Oncorhynchus mykiss*), as shown in studies by (Mongillo 2007; Ogunkalu 2019; Zheng et al. 2016).

## 1.13 Nanoparticle-Based Gene Therapy

Gene therapy is another well-practiced method used in the aquaculture sector (Xiang 2015). In contemporary times, the European Medicines Agency (EMA) has recommended the use of 'Clynav', a DNA vaccine for protecting Atlantic salmon against acute infections. In another report, an efficient carrier vehicle was designed using chitosan-dextran sulfate and silica nanoparticles to deliver dsRNA into *Penaeus monodon* post-larvae for silencing the Monodon baculovirus (MBV) structural gene, p74. This carrier system exhibited a significant survival rate (86.63%) after successful gene silencing in *P. monodon* (Ramesh Kumar et al. 2016).

## 1.14 Different Nanocarriers for Drug Delivery

**Chitosan**

Chitosan is a processed by-product of the shrimp industry and has been developed as a well-known polymer-based nanocarrier. Chitosan nanoparticles were orally delivered to black tiger shrimp (*Penaeus monodon*) loaded with a DNA construct comprising the VP28 gene of white spot syndrome virus (WSSV). A significant survivability rate was monitored in WSSV-challenged shrimp compared to the 100% mortality in the control group (Rajeshkumar et al. 2009). Chitosan microspheres exhibited positive trends as an orally delivered plasmid vaccine carrier for immunization in Japanese flounder (*Paralichthys olivaceus*) (Tian et al. 2008). Many studies have successfully implemented encapsulation and delivery by chitosan-based systems in aquaculture, viz. treatment methods have been developed against *Vibrio parahaemolyticus* in the blackhead seabream (*Acanthopagrus schlegelii*) (Li et al. 2013), *Philasterides dicentrarchi* in the turbot (*S. maximus*) (Leon-Rodrıguez et al. 2013), and *Vibrio anguillarum* in Asian sea bass (*Lates calcarifer*) (Kumar et al. 2008). Both DNA and RNA have been successfully encapsulated and delivered using this system. Studies by Ferosekahn et al. (2014) showed dietary RNA has been used in rohu (*Labeo rohita*), as well as inactive particles of the viral hemorrhagic septicemia virus (VHSV) in Japanese flounder (*Paralichthys olivaceus*), as reported by Kole et al. (2019). Recently, Kitiyodom et al. (2019) reported a chitosan-coated mucoadhesive nanovaccine with enhanced efficacy by immersion vaccination in tilapia (*Oreochromis* sp.) against columnaris disease. Additionally, chitosan-coated membrane vesicles (cMVs) derived from the pathogen *Piscirickettsia salmonis* were injected into adult zebrafish (*Danio rerio*), which conferred significant protection with an increased survival rate and induced an increased immune response (Tandberg et al. 2018).

## 1.15 Liposomes

Another commonly applied product is liposomes, which are artificially designed nanostructures comprising a lipid bilayer and have natural properties for application as carriers for drugs, vaccines, etc. It is nontoxic, biodegradable, and non-immunogenic, and several nanoliposomal products are approved by the FDA. Moreover, it serves as an efficient delivery system, as it possesses the capacity to control and target the release of hydrophobic and hydrophilic compounds that are easily permeable through cell membranes. Nanoliposomal encapsulated Omega-3 fatty acids synthesized from seafood and fish oil are used in food fortification for efficient delivery and protection (Hadian 2016). It is also used to deliver important vitamins like folic acid via a dermal route that paved the way for delivering nutraceuticals (Kapoor et al. 2018). Liposome-encapsulated *Aeromonas salmonicida* antigen exhibited effective immunization in carp (*Cyprinus carpio*) (Irie et al. 2005). Similarly, an immunostimulant cocktail was applied as a nonspecific nanocarrier of the vaccine to multiple fish species. A liposomal lipopolysaccharide (LPS)-dsRNA cocktail was observed entering into the hepatocytes of zebrafish and trout macrophage plasma membranes, triggering pro-inflammatory and antiviral reactivity (Ruyra et al. 2013). Positive results were recorded after encapsulating curcumin into liposomes, prepared from natural sources (salmon lecithin), and examining them under primary cortical neuron culture (Hasan et al. 2018).

## 1.16 Poly (Lactic-*Co*-Glycolic) Acid (PLGA)

Polymeric nanoparticles of poly (lactic-*co*-glycolic) acid (PLGA) are examples of well-established drug delivery vehicles having adjuvant properties. It was found that an antigenic (TNPLPH) and immunostimulant (β-glucan) candidate encapsulated within the PLGA nanoparticles had the potentiality to enhance the expression of pro-inflammatory markers in salmon (Fredriksen et al. 2011). PLGA nanoparticles have also been monitored as a potential carrier of DNA vaccine for oral immunization against the lymphocystis disease virus, as evaluated in Japanese flounder (*Paralichthys olivaceus*) (Tian and Yu 2011). Surfactant-free PLGA nanoparticles displayed optimal biocompatibility and targeted dendritic cells in adult zebrafish after mucosal administration. This experiment opens a powerful platform for mucosal vaccine delivery in aquaculture (Resseguier et al. 2017). Shah and Mraz (2019) showed the use of encapsulation and delivery of different compounds in fish. For example, a DNA vaccine encapsulated in PLGA showed an improved immunological response against lymphocystis (Tian and Yu 2011). Additionally, Behera et al. (2010) showed the potential of PLGA encapsulated antigens from *Aeromonas hydrophila*, used as a vaccine in Rohu, to produce an enhanced immune-stimulatory and antibody response, both at 21 and 42 days post-immunization. Similarly, Yun et al. (2017) synthesized PLGA microparticles to carry *A. hydrophila* cells that were previously killed by formalin treatment as a way to deliver antigens to pond loaches (*Misgurnus anguillicaudatus*) and common carps (*C. carpio*). Upon being challenged with *A. hydrophila*, the fishes were treated with PLGA-*A. hydrophila*, which showed a higher survival rate with increased innate and adaptive immune responses than the group treated just with the inactivated *A. hydrophila*, which could potentially induce higher and more lasting immune responses than the antigens alone. Additionally, mass vaccination can be achieved using nanocapsules resistant to digestion and degradation. Both oral administration and site-specific release of the active agent will reduce the effort and cost related to disease management, leading to more sustainable practices (Rather et al. 2011).

## 1.17 Solid Lipid Nanoparticles

Solid lipid nanoparticles (SLNs) are investigated due to their potential topical delivery and controlled release in the skin. Pandita et al. (2011) investigated the potential applications of solid lipid nanoparticles (SLNs) in improving the oral bioavailability of paclitaxel, a hydrophobic drug. Experiments were conducted to evaluate the delivery of solid lipid nanoparticles (SLN) encapsulating potential antioxidant glutathione (GSH) on immunocompetent fish cells of a seawater teleost, the gilthead seabream (*S. aurata L.*). It was observed that the phagocytic function of HK (head kidney) leucocytes incubated with encapsulated glutathione was significantly enhanced. A better *in vitro* antioxidant efficacy of SLN was also monitored (Trapania et al. 2016). Fin rot, dropsy, black and white spot disease, gill rot, columnaris, anchor worms, hole in the head, and *Argulus* infestations are some of the important external fish diseases that can be administered for recovery.

## 1.18 Microbial Disinfection

Nanoparticles synthesized from silver, titanium, and copper, among others, have been used for disease prevention and treatment. They have different modes of action against bacteria, and the most important function is acting on the cell membrane and cell wall by adhering to the cell wall through electrostatic interaction and disrupting the stability of the membrane. They can also disrupt the ion transport chain by interacting with ions and ion channels (Fajardo et al. 2022). It also generates double-strand breaks in the DNA, interfering with the ribosome assembly and enzymatic activity via electrostatic interactions. They are also known to trigger a higher oxidative stress state by increasing the number of reactive oxygen species (ROS) that can damage proteins, lipids, other cellular contents, and DNA. Colloidal silver nanoparticles are one of the most widely used nanotechnology products against a wide spectrum of pathogens, including viruses, parasites, fungi, and bacteria. Silver is considered to be one of the most effective nanomaterials among the oligo-dynamic metallic nanoparticles due to its wide-range effects on different microbial species, different application forms, crystallographic structure, high surface exposure compared to volume, and compatibility with several compounds, making it a go-to use candidate for any treatment (Nangmenyi and Economy 2014). It triggers the oxidation of DNA and proteins with highly damaging effects, thus exerting antimicrobial activity. For example, silver nanoparticles are effective against methicillin-resistant *S. aureus* (Jeong et al. 2005). Eliminating microbes using visible light photocatalysts mediated by nanoparticles of metal oxides, as well as nano-porous fibres and foams, were also found to be effective (Li et al. 2014). Their effect is not limited to microbial elimination but also includes the elimination of organic pollutants from the pharmaceutical and cosmetics industries (Chena and Yadab 2011). Titanium dioxide nanoparticles, for example, have a strong anti-bacterial activity and are capable of destroying fish pathogens *in vitro* (Cheng et al. 2009), and their actions have been confirmed as a strong immune modulator of fish neutrophil function (Jovanovic 2011). NanoCheck, a commercially available nanotechnology product that is currently used for cleaning fishponds with lanthanum-based particles (40 nm in size), functions by inhibiting algal growth by absorbing phosphates from the water (Rather et al. 2011). Korni and Khalil (2017) showed that ginger-derived nanoparticles control the infection by motile *Aeromonas septicaemia* in the Asian carp fingerlings. Rather et al. (2017) reported that silver nanoparticles synthesized by *Azadirachta indica* (neem) showed antibacterial and immunomodulatory activity in mrigal (*Cirrhinus mrigala*) fingerlings when challenged with *A. hydrophila*. More recently, Erdem et al. (2018) demonstrated the antibacterial efficacy of silver nanoparticles synthesized by *Aeromonas sobria* against *A. hydrophila*. This raised the possibility to improve sanitary management through the use of more eco-friendly and economically based antimicrobial agents (Shah and Mraz 2019). Recent studies by Kepiro et al. (2020) explain the conceptual design of protein pseudocapsids exerting broad-spectrum antimicrobial activity. In contrast to conventional antibiotics, these pseudocapsids are highly effective against diverse bacterial strains. This method can eliminate antibiotic-resistant bacteria *in vivo* without causing toxicity. Pseudocapsids, which are nothing but an icosahedral structure that is polymorphic in size, but not in shape, are available in both D and L epimeric forms. It has to be noted that pseudocapsids can cause fast and irreversible harm to bacterial cells.

## 1.19 Conclusions and Future Prospects

There are some significant disadvantages to the classical methods of treatment and remedy that have been observed. But those classical approaches have given a path for modern technology, without which science can't flourish. One such field is nanotechnology, which has given rise to numerous options with its wide field of application. Nowadays, disease in the aquaculture sector is bringing huge losses, and methods are being developed for rapid diagnosis and treatment. One such method is nanovaccination, which is a new attempt to enhance the immunogenicity of vaccines using nanoparticles as carriers and/or adjuvants. The immune system can very well be triggered because of the similar scale (size) between the nanoparticles and the pathogens, resulting in triggered cellular and humoral immunity responses. Thus, using nanotechnology in aquaculture has become a comprehensive tool for solving a lot of problems, not only in disease diagnosis and treatment but also in water quality control, fish nutrition, environmental management, etc. Currently, various nanotechnological applications have been implemented to improve the aquaculture industry, which could play an important role in the development and sustainability of this industry in the future. Nowadays, there are many potential applications for nanomaterials in the fisheries and aquaculture industries. Some of the most promising areas in this field are applications related to fish health management, nanoscale ingredient incorporation, the use of nanotechnology in aquaculture feeds and food packaging, as well as applications linked to value-added products, stress reduction, and health management. Sustainable development of nanotechnology in the fisheries and aquaculture industries will require a comprehensive assessment of its potential negative impacts; therefore, a careful analysis of the life cycle and shelf life of new nanomaterials, combined with an assessment of potential health and environmental risks, including exposure, release, and deposition, must be performed. However, considering the negative sides of this method, we should not refrain from trying to implement nanotechnological applications in the aquaculture sector, which should be linked to careful monitoring and controlled use, thus favouring efforts to minimize risks and maximize benefits. In the future, nanoparticles can be used in the detection of different diseases caused by bacteria as well as viruses in fishes (Tables 1.1 and 1.2).

**Table 1.1** Merits and demerits of nanoparticles

| Type of nanoparticles | Merits | Demerits |
|---|---|---|
| Polymeric nanoparticles | Better immunogenicity can be obtained by easy modification of surface properties, biodegradable and targeted antigen delivery | Low aqueous solubility and synthesis require the use of organic solvents, low antigen loading, premature release of antigens, insufficient antigen protection |
| Inorganic nanoparticles | Easy to modify, less chances of premature release, and better protection of adsorbed antigens | Low aqueous solubility and low biodegradability |
| Nanoliposomes | Possess intrinsic adjuvant properties, accommodate both hydrophilic and lipophilic antigens and relatively stable in gastrointestinal fluids when modified | Low mucus penetration, limited antigen loading, and poor gastrointestinal stability of naked liposomes |
| Virus-like particles (VLP's) | Possess self-adjuvant properties, mimic original virus and high gastrointestinal stability | Lack of reproducibility |
| Nanoemulsions | Possess self-adjuvant properties, encapsulate both hydrophilic and lipophilic antigens | Premature release of antigens and poor gastrointestinal stability |

**Table 1.2** Progress in nanoparticle-based fish vaccine delivery

| Nanoparticle | Pathogen | Fish species | Mode of administration | Vaccine formulations with nanoparticles | References |
|---|---|---|---|---|---|
| Chitosan | VHSV | *Danio rerio* | IP | NPrgpG pICrgpG | Kavaliauskis et al. (2016) |
| Chitosan | ISAV | *Salmo salar* | Oral | NP-V/ NP-Ad+NP-V | Aravena et al. (2013) |
| Chitosan/ TPP | Nodavirus | *Lates calcarifer* | Oral | pFNCPE42-CS/TPP | Vimal et al. (2014) |
| Chitosan | *Vibrio anguillarum (Listonella)* | *Lates calcarifer* | Oral | Chitosan-pVAOMP38 | Rajesh et al. (2008) |
| PLGA | *Aeromonas hydrophila* | *Labeo rohita* | Oral | Np-rOmpW (HiAg) 8 μg/g Np-rOmpW (LoAg) 4 μg/g | Dubey et al. (2016) |
| PLGA, PLA | *Aeromonas hydrophila* | *Labeo rohita* | IP | PLGA-Omp | Rauta and Nayak 2015 |
| PLGA | IPNV | *Salmo salar* | IP | PLGA nanoparticle-TA PLGA nanoparticle-PT | Munangandu et al. (2012) |

(continued)

**Table 1.2** (continued)

| Nanoparticle | Pathogen | Fish species | Mode of administration | Vaccine formulations with nanoparticles | References |
|---|---|---|---|---|---|
| PLGA | IHNV | *Oncorhynchus mykiss* | Oral | High dose PLGA(10WPV) Low dose PLGA-pCDNA-G high dose PLGA-pCDNA-G high dose PLGA | Adomako et al. (2012) |
| PLGA, PLA | – | *Salmo salar* | IP | PLGA-NP50L PLGA-NP50H PLGA-NP75H PLA-NP100L | Fredriksen and Grip (2012) |
| PLGA | LCDV | *Paralichthys olivaceus* | Oral | pEGFP-N2-MCP PLGA PLGA- pEGFP-N2-MCP | Tian and Yu (2011) |
| PLGA | – | *Salmo* saliar | IP | NP NP/TNP-LPH, NP/β glucan, NP/ TNP-LPH/β glucan | Fredriksen et al. (2011) |
| Liposome | *Vibrio harveyi* | *Epinephelus bruneus* | IP | Liposome-*V. harveyi* | Harikrishnan et al. (2012) |
| Liposome | *Aeromonas salmonicida* | *Cyprinus carpio* | Oral | Liposome-T1031 | Irie et al. (2005) |
| Carbon nanotubes | GCRV | *Ctenopharyngon idellus* | IM Bath | SWCNTs-pEGFP-vp5 1 μg SWCNTs-pEGFP-vp5 2.5 μg SWCNTs-pEGFP-vp5 5 μg SWCNTs-pEGFP-vp5 1 mgL-1 SWCNTspEGFP-vp5 10 mg L-1 SWCNTs-pEGFP-vp5 20 mg L-1 | Wang et al. (2015) |

## References

Adams A, Thompson KD (2006) Biotechnology offers revolution to fish health management. Trends Biotechnol 24:201–205

Adams A, Thompson KD (2008) Recent applications of biotechnology to novel diagnostics for aquatic animals. Rev Sci Tech Off Int Epiz 27:197–209

Adomako M, St-Hilaire S, Zheng Y, Eley J, Marcum RD (2012) Oral DNA vaccination of rainbow trout, Oncorhynchus mykiss (Walbaum), against infectious haematopoietic necrosis virus using PLGA [Poly (D, L-Lactic- Co-Glycolic Acid)] nanoparticles. J Fish Dis 35(3):203–214

Altenburger R, Brack W, Burgess RM, Busch W, Escher BI, Focks A et al (2019) Future water quality monitoring: improving the balance between exposure and toxicity assessments of real-world pollutant mixtures. Environ Sci Eur 31:1–17

Aravena AR, Sandino AM, Spencer E (2013) Nanoparticles and microparticles of polymers and polysaccharides to administer fish vaccines. Biol Res 46(4):407–419

Arfat YA, Benjakul S, Prodpran T, Sumpavapol P, Songtipya P (2014) Properties and antimicrobial activity of fish protein isolate/fish skin gelatin film containing basil leaf essential oil and zinc oxide nanoparticles. Food Hydrocoll 41:265–273

Assefa A, Abunna F (2018) Maintenance of fish health in aquaculture: review of epidemiological approaches for prevention and control of infectious disease of fish. Vet Med Int 5432497. https://doi.org/10.1155/2018/5432497

Bacchetta C, Ale A, Simoniello MF, Gervasio S, Davico C, Rossi AS et al (2017) Genotoxicity and oxidative stress in fish after a short-term exposure to silver nanoparticles. Ecol Indic 76:230–239

Behera T, Nanda PK, Mohanty C, Mohapatra D, Swain P, Das BK, Routray P, Mishra BK, Sahoo SK (2010) Parenteral immunization of fish, *Labeo rohita* with Poly D, L-lactide-co-glycolic acid (PLGA) encapsulated antigen microparticles promotes innate and adaptive immune responses. Fish Shellfish Immunol 28(2):320–325. https://doi.org/10.1016/j.fsi.2009.11.009

Bhattacharyya A, Reddy J, Hasan MM, Adeyemi MM, Marye RR (2015) Nanotechnology-a unique future technology in aquaculture for the food security. Int J Bioassays 4(7):4115–4126

Brisebois P, Siaj M (2019) Harvesting graphene oxide-years: 1859 to 2019: a review of its structure, synthesis, properties and exfoliation. J Mater Chem C. https://doi.org/10.1039/c9tc03251g

Chen B, Zou L, Wu Z, Sun M (2016) The application of quantum dots in aquaculture pollution detection. Toxicol Environ Chem 98(3–4):385–394. https://doi.org/10.1080/02772248.2015.1123482

Chena H, Yadab R (2011) Nanotechnologies in agriculture: new tools for sustainable development. Trends Food Sci Technol 22(11):585–594. https://doi.org/10.1016/j.tifs.2011.09.00

Cheng TC, Yao KS, Yeh N, Chang CI, Hsu HC, Chien YT, Chang CY (2009) Visible light activated bactericidal effect of TiO2/Fe3O4 magnetic particles on fish pathogens. Surf Coat Technol 204(6–7):1141–1144. https://doi.org/10.1016/j.surfcoat.2009.06.050

Crane M, Hyatt A (2011) Viruses of fish: an overview of significant pathogens. Virus 3:2025–2046

Cunningham CO (2004) Use of molecular diagnostic tests in disease control: making the leap from laboratory to fi eld application. In: Leung K-Y (ed) Current trends in the study of bacterial and viral fish and shrimp diseases, molecular aspects of fish and marine biology, vol 3. World Scientific, Singapore, pp 292–312

Dubey S, Avadhani K, Mutalik S, Sivadasan SM, Maiti B (2016) Aeromonas hydrophila OmpW PLGA nanoparticle oral vaccine shows a dose-dependent protective immunity in rohu (Labeo rohita). Vaccine 4(2):21

Erdem B, Dayangac A, Kıray E, Duygu D (2018) Biosynthesis of silver nanoparticles from *Aeromonas sobria* and antibacterial activity against fish pathogens. Int J Environ Sci Technol 16(9):1–6. https://doi.org/10.1007/s13762-018-1944-z

Fajardo C, Martinez G, Rodriguez J Blasco JM Mancera B Thomas MDD (2022). Nanotechnology in aquaculture: applications, perspectives and regulatory challenges. Aquac Fish 7(2): 185–200. ISSN 2468-550X. https://doi.org/10.1016/j.aaf.2021.12.006

FAO (2018) The state of world fisheries and aquaculture 2018 – meeting the sustainable development goals. Food & Agriculture Organization, Rome. Retrieved from http://www.fao.org/3/i9540en/i9540en.pdf

Ferosekhan S, Gupta S, Singh A, Rather M, Kumari R, Kothari D, Pal KA, Jadhao SB (2014) RNA-loaded chitosan nanoparticles for enhanced growth, immunostimulation and disease resistance in fish. Curr Nanosci 10(3):453–464. https://doi.org/10.2174/1573413710666140115220300

Firdaus-Nawi M, Saad M (2016) Major components of fish immunity: a review. Pertanika J Trop Agric Sci 39(4):393–420

Fredriksen BN, Grip J (2012) PLGA/PLA micro-and nanoparticle formulations serve as antigen depots and induce elevated humoral responses after immunization of Atlantic salmon (Salmo salar L.). Vaccine 30(3):656–667

Fredriksen BN, Saevareid K, McAuley L, Lane ME, Bogwald J et al (2011) Early immune responses in Atlantic salmon (*Salmo salar* L.) after immunization with PLGA nanoparticles loaded with a model antigen and β-glucan. Vaccine 29(46):8338–8349

Gao J, Lin, Wei A, Sepulveda (2017) Protein corona analysis of silver nanoparticles exposed to fish plasma. Environ Sci Technol Lett 4(5):174–179

Hadian Z (2016) A review of nanoliposomal delivery system for stabilization of bioactive Omega-3 fatty acids. Electron Phys. https://doi.org/10.19082/1776

Handy RD (2012) FSBI briefing paper: nanotechnology in fisheries and aquaculture. Fisheries Society of the British Isles

Harikrishnan R, Kim JS, Balasundaram C, Heo MS (2012) Vaccination effect of liposomes entrapped whole cell bacterial vaccine on immune response and disease protection in Epinephelus bruneus against Vibrio harveyi. Aquaculture 342–343:69–74

Hasan M, Latifi S, Kahn CJF, Tamayol A, Habibey R, Passeri E, Linder M, Arab- Tehrany E (2018) The positive role of curcumin-loaded salmon nanoliposomes on the culture of primary cortical neurons. Mar Drugs 16:218

Herbeck LS, Unger D, Wu Y, Jennerjahn TC (2013) Effluent, nutrient and organic matter export from shrimp and fish ponds causing eutrophication in coastal and back-reef waters of NE Hainan, tropical China. Cont Shelf Res 57:92–104

Husain S, Verma SK, Yasin D, Hemlataa, Rizvi MMA, Fatma T (2021) Facile green bio- fabricated silver nanoparticles from microchaete infer dose-dependent antioxidant and anti-proliferative activity to mediate cellular apoptosis. Bioorg Chem 107:104535

Irie T, Watarai S, Iwasaki T, Kodama H (2005) Protection against experimental Aeromonas salmonicida infection in carp by oral immunisation with bacterial antigen entrapped liposomes. Fish Shellfish Immunol 18(3):235–242. https://doi.org/10.1016/j.fsi.2004.07.006

Jeevanandam J, Barhoum A, Chan YS, Dufresne A, Danquah MK (2018) Review on nanoparticles and nanostructured materials: history, sources, toxicity and regulations. Beilstein J Nanotechnol 9:1050–1074

Jeong SH, Hwang YH, Yi SC (2005) Antibacterial properties of padded PP/PE nonwovens incorporating nano-sized silver colloids. J Mater Sci 40(20):5413–5418. https://doi.org/10.1007/s10853-005-4340-2

Jovanovic B (2011) Immunotoxicology of titanium dioxide and hydroxylated fullerenes engineered nanoparticles in fish models. Graduate theses and dissertations Iowa State University Capstones, Theses and Dissertations: 1–176

Ju-Nam Y, Lead JR (2008) Manufactured nanoparticles: an overview of their chemistry, interactions and potential environmental implications. Sci Total Environ 400:396–414

Kapoor MS, D'Souza A, Aibani N, Nair SS, Sandbhor P, Banerjee R (2018) Stable liposome in cosmetic platforms for transdermal folic acid delivery for fortification and treatment of micronutrient deficiencies. Sci Rep 8:1–12

Karunasagar I, Nayak BB, Karunasagar I (1997) Rapid detection of Vibrio parahaemolyticus from fi sh by polymerase chain reaction. In: Flegel TW, MacRae LH (eds) Diseases in Asian aquaculture III. Asian Fisheries Society, Manila, pp 119–122

Kaul S, Gulati N, Verma D, Mukherjee S, Nagaich U (2018) Role of nanotechnology in cosmeceuticals: a review of recent advances. J Pharma 3420204. https://doi.org/10.1155/2018/3420204

Kavaliauskis A, Arnemo M, Speth M, Lagos L, Rishovd AL et al (2016) Protective effect of recombinant VHSV-G vaccine using poly (I:C) loaded nanoparticles as an adjuvant in zebrafish (Danio rerio) infection model. Dev Comp Immunol 61:248–257

Kepiro IE, Marzuoli I, Hammond K, Ba X, Lewis H, Shaw M, Gunnoo SB, De Santis E, Kapinska U, Pagliara S, Holmes MA, Lorenz CD, Hoogenboom BW, Fraternali F, Ryadnov MG (2020) Engineering chirally blind protein pseudocapsids into antibacterial persisters. ACS Nano 14(2):1609–1622. https://doi.org/10.1021/acsnano.9b06814

Khan I, Saeed K, Khan I (2019) Nanoparticles: properties, applications and toxicities. Arab J Chem 12:908–931

Kitiyodom S, Yata T, Yostawornkul J, Kaewmalun S, Nittayasut N, Suktham K, Surassmo S, Namdee K, Rodkhum C, Pirarat N (2019) Enhanced efficacy of immersion vaccination in tilapia against columnaris disease by chitosan-coated "pathogen-like" mucoadhesive nanovaccines. Fish Shellfish Immunol 95:213–219. https://doi.org/10.1016/j.fsi.2019.09.064

Kole S, Qadiri SSN, Shin SM, Kim WS, Lee J, Jung SJ (2019) Nanoencapsulation of inactivated-viral vaccine using chitosan nanoparticles: evaluation of its protective efficacy and immune modulatory effects in olive flounder (*Paralichthys olivaceus*) against viral haemorrhagic septicaemia virus (VHSV) infection. Fish Shellfish Immunol 91:136–147. https://doi.org/10.1016/j.fsi.2019.05.017

Korni FMM, Khalil F (2017) Effect of ginger and its nanoparticles on growth performance, cognition capability, immunity and prevention of motile Aeromonas septicaemia in Cyprinus carpio fingerlings. Aquac Nutr 23(6):1492–1499. https://doi.org/10.1111/anu.12526

Kumar SR, Ahmed VI, Parameswaran V, Sudhakaran R, Babu VS, Hameed AS (2008) Potential use of chitosan nanoparticles for oral delivery of DNA vaccine in Asian sea bass (*Lates calcarifer*) to protect from *vibrio* (*Listonella*) *anguillarum*. Fish Shellfish Immunol 25(1–2):47–56. https://doi.org/10.1016/j.fsi.2007.12.004

Kumar P, Huo P, Zhang R, Liu B (2019) Antibacterial properties of graphene-based nanomaterials. Nano. https://doi.org/10.3390/nano9050737

Kumari P, Panda PM, Jha E, Kumari K, Nisha K, Mallick MA, Verma SK (2017) Mechanistic insight to ROS and apoptosis regulated cytotoxicity inferred by green synthesized CuO nanoparticles from *Calotropis gigantea* to embryonic zebrafish. Sci Rep 7:1–17

Leon-Rodrıguez L, Luzardo-Alvarez A, Blanco-Mendez J, Lamas J, Leiro J (2013) Biodegradable microparticles covalently linked to surface antigens of the scuticociliate parasite P. dicentrarchi promote innate immune responses in vitro. Fish Shellfish Immunol 34(1):236–243. https://doi.org/10.1016/j.fsi.2012.10.029

Li L, Lin SL, Deng L, Liu ZG (2013) Potential use of chitosan nanoparticles for oral delivery of DNA vaccine in black seabream *Acanthopagrus schlegelii* Bleeker to protect from Vibrio parahaemolyticus. J Fish Dis 36(12):987–995. https://doi.org/10.1111/jfd.12032

Li Q, Wu P, Shang JK (2014) Nanostructured visible-light photocatalysts for water purification. In: Street A, Sustich R, Duncan J, Savage N (eds) Nanotechnology applications for clean water: solutions for improving water quality. William Andrew, Norwich, pp 297–317. https://doi.org/10.1016/B978-1-4557-3116-9.00019-6

Mahanty A, Bosu R, Panda P, Netam SP, Sarkar B (2013a) Microwave assisted rapid combinatorial synthesis of silver nanoparticles using *E.coli* culture supernatant. Int J Pharma Biosci 4:1030–1035

Mahanty A, Mishra S, Bosu R, Maurya UK, Netam SP, Sarakar B (2013b) Phytoextracts-synthesized silver nanoparticles inhibit bacterial fish pathogen Aeromonas hydrophila. Ind J Microbiol 53:438–446

Mamo T, Poland GA (2012) Nanovaccinology: the next generation of vaccines meets 21st century materials science and engineering. Vaccine 30:6609–6611

Matteucci F, Giannantonio R, Calabi F, Agostiano A, Gigli G, Rossi M (2018) Deployment and exploitation of nanotechnology nanomaterials and nanomedicine. In: AIP conference proceedings. https://doi.org/10.1063/1.5047755

Mohamed S, Mona S, Magdy EM, Mansour EM (2015) Recent progress in applications of nanoparticles in fish medicine: a review. Nanomed Nanotechnol Biol Med. https://doi.org/10.1016/j.nano.2015.11.005

Mongillo JF (2007) Nanotechnology 101. Greenwood Press, Westport

Munangandu HM, Fredriksen BN, Mutoloki S, Brudeseth B, Kuo TY et al (2012) Comparison of vaccine efficacy for different antigen delivery systems for infectious pancreatic necrosis virus vaccines in Atlantic salmon (Salmo salar L.) in a cohabitation challenge model. Vaccine 30(27):4007–4016

Nangmenyi G, Economy J (2014) Nanometallic particles for oligodynamic microbial disinfection. In: Street A, Sustich R, Duncan J, Savage N (eds) Nanotechnology applications for clean water: solutions for improving water quality. William Andrew, Norwich, pp 283–295

Natarajan A, Shalini Devi KS, Raja S, Senthil Kumar A (2017) An elegant analysis of white spot syndrome virus using a graphene oxide/methylene blue based electrochemical immunosensor platform. Sci Rep. https://doi.org/10.1038/srep46169

Naylor RL, Goldburg RJ, Primavera JH, Kautsky N, Beveridge MCM, Clay J et al (2006) Effect of aquaculture on world fish supplies. Nature 405:1017–1024

Ogunkalu OA (2019) Utilization of nanotechnology in aquaculture and seafood sectors. EJFST 19:26–33

Pandita D, Ahuja A, Lather V, Benjamin B, Dutta T, Velpandian T (2011) Development of lipid-based nanoparticles for enhancing the oral bioavailability of Paclitaxel. AAPS Pharm Sci Technol 12:712–722

Patel M, Kumar R, Kishor K, Mlsna T, Pittman CU, Mohan D (2019) Pharmaceuticals of emerging concern in aquatic systems: chemistry, occurrence, effects, and removal methods. Chem Rev 119:3510–3673

Patolsky F, Zheng GF, Hayden O, Lakadamyali M, Zhuang XW, Lieber CM (2004) Electrical detection of single viruses. Proc Natl Acad Sci 101:14017–14022. https://doi.org/10.1073/pnas.0406159101

Patra JK, Das G, Fraceto LF, Campos E, Rodriguez-Torres M, Acosta-Torres LS et al (2018) Nano based drug delivery systems: recent developments and future prospects. J Nanobiotechnol 16:71. https://doi.org/10.1186/s12951-018-0392-8

Puttinaowarat S, Thompson KD, Adams A (2000) Mycobacteriosis: detection and identification of aquatic mycobacterium species. Fish Vet J 5:6–21

Rajesh SK, Ahmed VPI, Parameswaran V, Sudhakaran R, Babu VS et al (2008) Potential of chitosan nanoparticles for oral delivery of DNA vaccine in Asian sea bass (*Lates calcarifer*) to protect from vibrio (Listonella) anguillarum. Fish Shellfish Immunol 25(1–2):47–56

Rajeshkumar S, Venkatesan C, Sarathi M, Sarathbabu V, Thomas J, Anver Basha K, Hameed ASS (2009) Oral delivery of DNA construct using chitosan nanoparticles to protect the shrimp from white spot syndrome virus (WSSV). Fish Shellfish Immunol 26:429–437. https://doi.org/10.1016/j.fsi.2009.01.003

Rajkumar KS, Kanipandian N, Ramasamy T (2016) Toxicity assessment on haemotology, biochemical and histopathological alterations of silver nanoparticles-exposed freshwater fish Labeo rohita. Appl Nanosci 6(1):19–29

Ramesh Kumar D, Elumalai R, Raichur AM, Sanjuktha M, Rajan JJ, Alavandi SV, Vijayan KK, Poornima M, Santiago TC (2016) Development of antiviral gene therapy for monodon baculovirus using dsRNA loaded chitosan-dextran sulfate nanocapsule delivery system in Penaeus monodon post-larvae. Antivir Res 131:124–130

Rather MA, Sharma R, Aklakur M, Ahmad S, Kumar N, Khan M, Ramya VL (2011) Nanotechnology: a novel tool for aquaculture and fisheries development. A prospective mini-review. Fish Aquac J FAJ 16

Rather MA, Bhat IA, Sharma N, Gora A, Ganie PA, Sharma R (2017) Synthesis and characterization of Azadirachta indica constructed silver nanoparticles and their immunomodulatory activity in fish. Aquac Res 48:3742–3754

Rauta PR, Nayak B (2015) Parenteral immunization of PLA/PLGA nanoparticle encapsulating outer membrane protein (Omp) from *Aeromonas hydrophila*. Evaluation of immunostimulatory action in *Labeo rohita* (rohu). Fish Shellfish Immunol 44(1):287–294

Resseguier J, Delaune E, Coolen AL, Levraud JP, Boudinot P, Guellec DL, Verrier B (2017) Specific and efficient uptake of surfactant-free poly (Lactic Acid) nanovaccine vehicles by mucosal dendritic cells in adult zebrafish after bath immersion. Front Immunol. https://doi.org/10.3389/fimmu.2017.00190

Roy S, Kumar V, Barman D, Mandal S (2012) Nanotechnology and its application in fisheries and aquaculture. AQUA Int 20(6):2–3

Ruyra A, Cano-Sarabia M, MacKenzie SA, Maspoch D, Roher N (2013) A novel liposome-based nanocarrier loaded with an LPS-dsRNA cocktail for fish innate immune system stimulation. PLoS One 8:e76338

Sarkar B, Mahanty A, Gupta SK, Choudhury AR, Daware A, Bhattacharjee S (2022) Nanotechnology: a next-generation tool for sustainable aquaculture. Aquaculture 546:737330

Shah BR, Mraz J (2019) Advances in nanotechnology for sustainable aquaculture and fisheries. Rev Aquac 1–18. https://doi.org/10.1111/raq.12356

Siddiqi KS, Husen A, Rao R (2018) A review on biosynthesis of silver nanoparticles and their biocidal properties. J Nanobiotechnol 16:14

Stern ST, Martinez MN, Stevens DM (2016) When is it important to measure unbound drug in evaluating nanomedicine pharmacokinetics? Drug Metabol Disposal 44:1934–1939

Stone V, Nowack B, Baun A, van den Brink N, von der Kammer F, Dusinska M, Handy R, Hankin S, Hassellov M, Joner E, Fernandes TF (2010) Nanomaterials for environmental studies: classification, reference material issues, and strategies for physico-chemical characterisation. Sci Total Environ 408:1745–1754. https://doi.org/10.1016/j.scitotenv.2009.10.035

Tandberg J, Lagos L, Ropstad E, Smistad G, Hiorth M, Winther-Larsen HC (2018) The use of chitosan-coated membrane vesicles for immunization against salmonid rickettsial septicemia in an adult zebrafish model. Zebrafish 15(4):372–381. https://doi.org/10.1089/zeb.2017.1556

Thiruppathiraja C, Kumar S, Murugan V, Adaikkappan P, Sankaran K, Alagar M (2011) An enhanced immuno-dot blot assay for the detection of white spot syndrome virus in shrimp using antibody conjugated gold nanoparticles probe. Aquaculture 318:262–267

Thompson KD, Adams A (2004) Current trends in immunotherapy and vaccine development for bacterial diseases of fish. In: Leung K-Y (ed) Current trends in the study of bacterial and viral fish and shrimp diseases, Molecular aspects of fish and marine biology, vol 3. World Scientific, Singapore, pp 313–362

Tian J, Yu J (2011) Poly (lactic-co-glycolic acid) nanoparticles as candidate DNA vaccine carrier for oral immunization of Japanese flounder (*Paralichthys olivaceus*) against lymphocystis disease virus. Fish Shellfish Immunol 30:109–117. https://doi.org/10.1016/j.fsi.2010.09.016

Tian J, Yu J, Sun X (2008) Chitosan microspheres as candidate plasmid vaccine carrier for oral immunisation of Japanese flounder (*Paralichthys olivaceus*). Vet Immunol Immunopath 126:220–229

Toranzo AE, Magarinos TB, Romalde JL (2005) A review of the main bacterial fish diseases in mariculture systems. Aquaculture 246:37–61

Trapania A, Tripodob G, Mandracchiaa D, Cioffic N, Ditarantoc N, Cerezuela R, Esteband MA (2016) Glutathione loaded solid lipid nanoparticles: preparation and *in vitro* evaluation as delivery systems of the antioxidant peptide to immunocompetent fish cells. J Cell Biotechnol 2:1–14

Verma SK, Jha E, Sahoo B, Panda PK, Thirumurugan A, Parashare SKS, Suar M (2017) Mechanistic insight into the rapid one-step facile biofabrication of antibacterial silver nanoparticles from bacterial release and their biogenicity and concentration-dependent in vitro cytotoxicity to colon cells. RSC Adv 7:40034–40045

Verma SK, Nisha K, Panda PK, Patel P, Kumari P, Mallick MA, Sarkar B, Das B (2020) Green synthesized MgO nanoparticles infer biocompatibility by reducing in vivo molecular nanotoxicity in embryonic zebrafish through arginine interaction elicited apoptosis. Sci Total Environ 713:136521

Verma SK, Panda PK, Kumari P, Patel P, Arunima A, Jha E, Husain S, Prakash R, Hergenroder R, Mishra YK, Ahuja R, Varma RS, Suar M (2021) Determining factors for the nano-biocompatibility of cobalt oxide nanoparticles: proximal discrepancy in intrinsic atomic interactions at differential vicinage. Green Chem 23:3439–3458

Vevers WF, Jha AN (2008) Genotoxic and cytotoxic potential of titanium dioxide (TiO2) nanoparticles on fish cells in vitro. Ecotoxicology 17(5):410–420

Vimal S, Majeed AS, Nambi KSN, Madan N, Farook MA et al (2014) Delivery of DNA vaccine using chitosan-tripolyphosphate (CS/TPP) nanoparticles in Asian sea bass. *Lates calcarifer* (Bloch,1790) for protection against nodavirus infection. Aquaculture 420–421:240–246

Vinay TN, Bhat S, Choudhury TG, Paria A, Jung MH, Kallappa GS, Jung SJ (2018) Recent advances in application of nanoparticles in fish vaccine delivery. Rev Fish Sci Aquacult 26:29–41

Vinitnantharat S, Gravningen K, Greger E (1999) Fish vaccines. Adv Vet Med 41:539–550

Wang Y, Liu GL, Li DL, Ling F, Zhu B et al (2015) The protective immunity against grass carp reovirus in grass carp induced by a DNA vaccination using single-walled carbon nanotubes as delivery vehicles. Fish Shellfish Immunol 47(2):732–742

Wilson T, Carson J (2003) Development of sensitive, high-throughput one-tube RT-PCR-enzyme hybridisation assay to detect selected bacterial fish pathogens. Dis Aquatic Org 54:127–134

Xiang J (2015) Recent major advances of biotechnology and sustainable aquaculture in China. Curr Biotechnol 4:296–310

Yun S, Jun JW, Giri SS, Kim HJ, Chi C, Kim SG, Kim SW, Park SC (2017) Efficacy of PLGA microparticle-encapsulated formalin-killed Aeromonas hydrophila cells as a single-shot vaccine against A. hydrophila infection. Vaccine 35(32):3959–3965. https://doi.org/10.1016/j.vaccine.2017.06.005

Zhang JS, Li ZJ, Wen GL, Wang YL, Luo L, Zhang HJ, Dong HB (2016) Relationship between white spot syndrome virus (WSSV) loads and characterizations of water quality in *Litopenaeus vannamei* culture ponds during the tropical storm. Iran J Vet Res 17:210–214

Zhao L, Seth A, Wibowo N, Zhao C-X, Mitter N, Yu C, Anton PJ, Middelberg APJ (2014) Nanoparticle vaccines. Vaccine 32(3):327–337. https://doi.org/10.1016/j.vaccine.2013.11.069

Zheng F, Liu H, Sun X, Zhang Y, Zhang B (2016) Development of oral DNA vaccine based on chitosan nanoparticles for the immunization against reddish body iridovirus in turbots (Scophthalmus maximus). Aquaculture 452:263–271

# Chapter 2
# Nanotechnologies in Controlling Aquatic Diseases

**Haimanti Mondal and John Thomas**

## 2.1 Introduction

Aquaculture is the age-old practice of tending to confined water for growing aquatic organisms such as fish and shellfish and harvesting the production for human benefit. It is the human-controlled cultivation and harvest of freshwater and marine plants and animals. It includes fish farming, fish culture, mariculture, fish breeding, and ocean ranching. Throughout the world, aquaculture operations constitute an integral part of fisheries and aquatic resource management. Organisms as varied as trout, carp, and tuna (i.e., finfish), shrimps and oysters (i.e., shellfish), and seaweed are grown, using ponds, tanks, or nets, in salt, brackish, and fresh waters.

Aquaculture in India is one of the leisure activities among fishermen and farmers. With the passage of time, due to the limitations of the land-based food supply and the Malthusian fear, man has turned more seriously to water-based production systems. Earlier, it was easier to harvest wild fish stocks when compared to culturing fish, but now a stage of saturation in production through capture fisheries has been reached. Globally, annual water-based production has reached a plateau phase of 95–100 million tonnes during the current decade.

A vaccine is a biological preparation that improves immunity to a particular disease. The agent stimulates the body's immune system to recognize the agent as a foreign body. It destroys the foreign body and "reminisces" it so that the immune system can easily recognize and destroy any of these microorganisms that it later encounters. There are a lot of varieties of vaccines like DNA vaccines, recombinant vaccines, and many more. Whole-cell vaccine is a bacterial suspension of whole bacterial cells that have been killed. Whole-cell vaccine production is cheaper. It is

H. Mondal · J. Thomas (✉)
Centre for Nanobiotechnology, Vellore Institute of Technology (VIT), Vellore, TN, India
e-mail: john.thomas@vit.ac.in



V. K. Arivarasan et al. (eds.), *Nanotechnological Approaches to the Advancement of Innovations in Aquaculture*, Nanotechnology in the Life Sciences, https://doi.org/10.1007/978-3-031-15519-2_2

of two types: killed vaccine and attenuated vaccine. Though attenuated vaccine provides both cellular and humoral immunity, there is always a chance of infection in immunocompromised individuals and also a possibility of reversion to pathogenic forms. Thus, whole-cell killed vaccines are preferred as they are able to provide potent immunization. Whole-cell vaccines are preferred over other types because they are cheaper than others and also very effective. Though others have been found effective, their approval is still an issue, and also their production is difficult as in the case of the pertussis vaccine (Halperin et al. 1992). Though this is a human vaccine, this issue still affects other vaccines intended for other living forms.

Zhao et al. (2014), Pankhurst et al. (2003), and Tissot et al. (2008) reported that nanotechnology has made a huge progress in the field of biomedicine. In addition, the application has been increasing in the area of vaccinology, which gave rise to "nanovaccinology" (Mamo and Poland 2012; Zhao et al. 2014). Thus, the nanovaccines developed contain nanoparticles formulated with antigens either absorbed on or encapsulated within the surface against which an immune reaction is elicited (Gregory et al. 2013; Zaman et al. 2013).

In the past few decades, nanoparticles have been used as delivery systems and adjuvants in vaccines. Nanotechnology has formulated various efficient vaccine delivery systems that have helped in protecting the encapsulated antigens from the belligerent gastrointestinal environment. They also maintained the sustained release, which was inducing the immunostimulatory properties of the vaccine. Moreover, nanotechnology has been applied in the development of several fish vaccines for mass vaccination by either incorporating them in feed or administering them via immersion. Thus, nanovaccines can be a potential alternative and possible solution for injection-free mass vaccination and its applications in the aquaculture industry (Vinay et al. 2016).

Poly (lactic-co-glycolic acid) (PLGA) and polymeric chitosan are the most investigated nanoparticles in the fish vaccine research area (Mohamed et al. 2016). PLGA is a synthetic polymeric nanoparticle, whereas chitosan is a natural polymeric nanoparticle (Liang et al. 2014; Nirmal et al. 2014). Chitosan nanoparticles have been reported in the development of various fish vaccines. A study was reported on a nanoparticle-based oral vaccine against infectious salmon anemia virus (ISAV) that incorporated an alphavirus replicon as an adjuvant (Aravena et al. 2015).

## 2.2 Fish Vaccine in Aquaculture

The fish immune system is exposed to part of a pathogen or the entire pathogen (antigen), allowing time for the immune system to develop a response. Fish vaccines are classified as killed and modified live vaccines. Killed fish vaccines consist of killed (heat-killed/formalin-killed) pathogenic bacteria that stimulate the immune systems of the water-based production organisms and generate an immune response.

In order to prepare the bacterins, each bacterial isolate was inoculated separately into tryptic soy broth (TSB) and incubated for 24 h at 25 °C. Formalin (40% w/v) was added to the broth culture at a final concentration of 0.5% (V/V) and left for 48 h at room temperature. In the case of bivalent formalin-killed vaccine formulation, an equal portion of each bacterin was added to make one volume of the vaccine. Besides, the heat-killed vaccine was prepared by heating the broth culture for 30 min at 100 °C. The inactivated cells were counted with the hemocytometer ($1 \times 10^8$ cells/ml) for all the isolates. After that, the bacterins were tested for their sterility (free from the living cells) by streaking them onto trypticase soy agar, which showed no growth (Dehghani et al. 2012).

Pridgeon and Klesius (2010) reported that the modified live vaccines are comprised of live microorganisms that are grown in culture. They cannot cause significant disease. Live attenuated vaccines work by stimulating both cell-mediated and humoral immune responses and conferring protection for a long time.

Other than PLGA and chitosan, numerous other kinds of nanoparticles have been used in fish vaccine delivery including nanoliposomes, calcium phosphate, carbon nanotubes, immunostimulating complexes, and biodegradable polymers that have the potential to develop new vaccines against various fish pathogens (Vinay et al. 2017).

## 2.3 Types of Vaccines and Their Applications in Nanotechnology

### *2.3.1 Fish Vaccines*

With an increase in water-based products for human and animal consumption, a decline in fishery resources is also observed simultaneously. Hence, there is an additional need for alternate sources of water-based products, mainly through the aquaculture industry, which is claimed to be the fastest-growing segment of agriculture in the world (Hanfman 1993). The increased demand for aquatic animals, which is partly due to greater health awareness among consumers and the decline or stagnation of natural harvests, has largely contributed to this rapid growth.

Fish management, coupled with good hygiene practices, remains a major key factor in aquaculture-based production systems.

Recent researches are focusing on the importance of nanoparticles in terms of their application in the development of fish vaccines in aquaculture (Yildirimir et al. 2011). Nanoparticles are able to exhibit properties that are interesting as well as different compared to their parent compounds, including quantum size effects and increased relative surface area. Studies have proven that the application of nanoparticles has led to the enhancement of the stability, solubility, permeability, targeting, and biocompatibility of vaccines (You et al. 2012; Lai et al. 2013; Frohlich 2012; Doll et al. 2013).

#### 2.3.1.1 Viral Fish Vaccines

Benmansour and de Kinkelin (1997), Lopez Doriga et al. (2001), and Ronen et al. (2003) reported that most of the available virus vaccines for aquaculture nowadays are based on inactivated viruses or recombinant subunit proteins. Inactivated or killed viral vaccines are generally not successful unless delivered by injection. High doses are required to achieve protection. Live viral vaccines have been tested with good results in fish. The first viral vaccine for fish was against a carp rhabdovirus, causing spring viremia of carp (SVC). Commercially available vaccines for IPNV are based on either inactivated cell culture-propagated viruses or recombinant structural proteins. However, DNA vaccines encoding the same viral glycoproteins are remarkably efficacious. Indeed, these DNA vaccines are protective when used in small doses and efficacious as early as 4–8 days and for up to 2 years post vaccination (Corbeil et al. 2000).

A study was carried out to improve the prophylactic efficacy of immersion vaccines against fish viral diseases by constructing a targeted single-walled carbon nanotube (SWCNT)-based immersion vaccine delivery system "CNTs-M-VP7." The surface of this delivery system was modified with mannose to allow the targeting of antigen-presenting cells (APCs). This monosylated nanoparticle-based immersion vaccine was able to enter into the fish body through mucosal tissues like gill, skin, and intestine and later present to immune-associated tissues. They could trigger robust immune responses by inducing the maturation as well as the presenting process of the APCs (Zhu et al. 2020).

#### 2.3.1.2 Bacterial Fish Vaccines

The first commercially available bacterial vaccines were against enteric red mouth disease (ERM, yersiniosis) and vibriosis, introduced in the USA in the later 1970s (Evelyn 1997). These vaccines were based on inactivated whole cell formulations and were administered by immersion. Such vaccines have proven efficacious in preventing many of the major bacterial diseases. Vibriosis is an example of a disease against which the simple inactivated bacterin vaccine works well, but other bacteria have proven more difficult to control by vaccination.

Kuzyk et al. (2001) reported the development of new vaccines based on recombinant proteins. These new approaches might offer a solution for diseases where inactivated bacterins are inefficient, although the long-term performance of the vaccines remains to be documented.

The gram-negative bacteria were reported to hamper the growth of finfish in aquaculture. All gram-negative bacteria have surface-associated outer membrane proteins (OMPs), and they were known as potential vaccine candidates. A study revealed the applications of OMPs in designing certain vaccines based on subunit vaccines, DNA vaccines, chimeric proteins, and recombinant proteins as potential new-generation vaccine candidates for various bacterial pathogens in aquaculture. OMPs play a significant role in the adaptive responses of bacteria including iron acquisition, solute and ion uptake, bile salt resistance, antimicrobial resistance, serum resistance, and also maintaining the integrity as well as selective permeability of the bacterial cell membrane. OMPs that have been found conserved across serotypes were used as potential candidates in the design of vaccines (Maiti et al. 2020).

#### 2.3.1.3 Fish Vaccines Against Parasites

Parasitic diseases such as white spot disease, whirling disease, amoebic gill disease, proliferate kidney disease (PKD), and salmon live infestation create severe problems in fish farming, and no parasite vaccines are commercially available.

### 2.3.2 Need for New Fish Vaccines

Aasjord and Slinde (1994) and Mialhe et al. (1995) reported that outbreaks of infections can cause huge economic losses to fish farmers. There are no prophylactic or therapeutic measures available. Chemicals and antibiotics can be used to control bacterial and parasitic diseases, but these products often have undesirable side effects (Munn 1994).

In the case of viral infections that affect the water-based production systems of farms, there are no interventions available. Hence, a massive destruction of the infected stock is the only remedy. Current research in this field is focusing on disease prevention rather than treatment after infection. However, treatment can be opted for whenever the vaccine is unavailable. Vaccines provide herd immunity, but the currently available vaccines provide protection against bacterial diseases (Newman 1993; Schnick et al. 1997).

The high cost of new product development combined with the relatively small size of the industry and the low value of individual animals have largely contributed to this situation. There is a limited supply of many of the vaccines though tested and proven under laboratory conditions. This bias is due to the prohibitive cost of production, insufficient protection, or lack of safety (Leong et al. 1997; Munn 1994; Newman 1993). The lack of effective viral vaccines is one of the main problems in fish vaccinology.

### 2.3.3 DNA Vaccine

DNA vaccines are the most efficient vaccines established against viral diseases in fish to date only at an experimental level. DNA vaccines represent a powerful new approach to raising immune responses. The antigens are synthesized in transfected cells and obey the modification and antigen presentation rules of eukaryotic cells.

Very low levels of antigen (typically monogram levels) induce both antibody and cytolytic T-cell responses. Several reviews on DNA vaccines for fish are available (Anderson and Leong 2000; Heppell and Davis 2000; Jones 2001; Lorenzen et al. 2005).

Recently, research has diverted to the preparation of DNA vaccines that encode the pathogenic antigens and, when administered to fish, provide an immune response at an effective level (Boudinot et al. 1998; Lorenzen et al. 2000; Lorenzen et al. 2002; Mclauchlan et al. 2003; Pasnik and Smith 2005; Purcell et al. 2004; Vesely et al. 2004). Earlier, several research studies in fish involved the use of genes to study the magnitude of expression levels under different conditions with a focus on genes encoding proteins such as β-galactosidase, green fluorescent protein, and luciferase (Gomez Chiarri et al. 1996).

**Advantages of DNA Vaccines**

DNA vaccines offer several advantages to aquatic organisms over classical antigen vaccines (i.e., live attenuated, whole killed, and subunit vaccines). Practically. DNA vaccines are relatively inexpensive and easy to produce. Multivalent vaccines can also be easily prepared by mixing together different plasmids or including more than one antigen-encoding gene in a single vector for collinear expression, which will further reduce the cost of production. In addition, DNA is a very stable molecule at higher temperatures; therefore, shipment and storage need not be in a cold environment. All these factors contribute to making DNA vaccines very attractive for controlling fish diseases. DNA-based immunization also has immunological advantages over traditional methods of vaccination. It can induce strong and lasting humoral and cell-dependent immune responses without a boost, similar to that conferred by live vaccines, but without the risk of inadvertent infection (Davis and McCluskie 1999).

### 2.3.4 DNA Vaccines for Fish Pathogens in Nanotechnology

Gomez Chiarri et al. (1996) reported that one of the first bacterial fish pathogens for which DNA vaccines were tested was *Renibacterium salmoninarum*, the causative agent of bacterial kidney disease in salmon and trout, but no protective effect has been reported. A more generic approach has been attempted for *Piscirickettisa salmonis*, against which fish were vaccinated with a full expression library of plasmid DNA. A pathogen-specific antibody response was subsequently detected, but the level of protection was relatively low. Sommerset et al. (2005) reported that the viral hemorrhagic septicemia virus vaccine induced a high level of protection against Atlantic halibut nodavirus in turbot when the challenge was performed shortly after vaccination, thus demonstrating early protection and is not limited to rhabdovirus infections in salmonid.

The DNA vaccine encoding the outer membrane porin protein of *Vibrio anguillarum* in sea bass also provides a good immune response in sea bass (Rajesh Kumar et al. 2007).

#### 2.3.4.1 Oral Vaccine

Oral vaccines can be administered effectively to all fishes, regardless of their size. This method is a cheap way to either vaccinate or boost the immune status of any fish in any cultural environment. However, when administering oral vaccines to fishes, the attention is to be focused on overcoming degradation in the digestive environment (gut) of the fish so that an effective vaccine delivery is possible with an adequate immune response. Other delivery challenges include cost-effective production, shelf storage stability, and treatment to prevent leaching from feed upon contact with water.

An oral DNA vaccine was designed by loading the bacterial outer membrane protein K (ompK) gene of *Vibrio parahaemolyticus* onto the chitosan nanoparticles. They later elicited an immune response in black sea bream, *Acanthopagrus schlegelii*, against the pathogen *Vibrio parahaemolyticus* (Li et al. 2013). In another study, the pH-controlled release of dihydrolipoamide dehydrogenase (DLDH) antigens via a mesoporous silica nanoparticle (MSN) delivery system was used to develop an oral fish vaccine. The DLDH antigens of the bacteria, *Vibrio alginolyticus*, were loaded onto the MSN to design the vaccine delivery system. Moreover, hydroxypropyl methylcellulose phthalate (HP55) was coated to ensure the protection of the immunogen. They displayed the prepared MSN delivery system as a potential candidate carrier for fish vaccines through oral administration in aquaculture (Zhang et al. 2021).

## 2.4 Vibrosis: A Common Disease Pathogen in Aquaculture

A disease outbreak in aquaculture causes significant economic loss, but the use of antibiotics is not always preferable due to the development of drug-resistant strains. Antibiotic residue may also remain in the fishes, which can be toxic to them as well as humans. Sometimes, when the disease is not detected earlier, it finds its way into the human food chain, where the infection spreads to humans. Some of the diseases common in fisheries are furunculosis, dropsy, ergasilosis, lernaesosis, pike fly virus, viral hemorrhagic septicemia (VHS), and many more.

Vibriosis in aquaculture has been reported in various parts of the world. In India also, it is very much common. Thus, it has been a major concern. Vibriosis is caused by *Vibrio* spp. and has been a serious disease problem in prawns. Some of the *Vibrio* spp. include *V. harveyi*, *V. parahaemolyticus*, *V. anguillarum*, *V. vulnificus*, and *V. splendidus*, and some of the diseases they cause are tail necrosis, shell disease, red disease, loose shell syndrome (LSS), and white gut disease (WGD).

A combination of one or more of these can cause these diseases, and the virulence of the disease depends on the strain, its source, and its environment. The severity of infection depends on the species and strain of ***Vibrio*** involved, the stage of development and age of the prawn, and the ambient environmental conditions (Jayasree et al. 2006). Mass mortality is observed in different parts of the world in hatcheries and growing ponds of prawns (Couch 1978; Overstreet 1978; Lightner 1983, 1985, 1988, 1996; Sindermann 1990; Ruangpan and Kitao 1992; Chen et al. 1992; Yang et al. 1992; de la Pena et al. 1993; Jiravanichpaisal et al. 1994; Mohney et al. 1994; Lavilla-Pitogo and de la Pena 1998; Lavilla-Pitago et al. 1998). Over a dozen species have been isolated in case of implicated diseases (Overstreet 1978; Lightner 1988, 1996; Sindermann 1990). Vibriosis is also rampant in the Indian region, and infection from luminous vibriosis has caused many hatcheries to shut down (Couch 1978).

*Vibrio parahaemolyticus*, a marine bacterium, causes food-borne disease in humans, resulting in gastroenteritis. This is due to the possession of hemolysin genes (tdh. Trh. or both) that favor the progression of the disease (Kim et al. 1999). The mode of infection of *Vibrio parahaemolyticus* in fish is mainly via penetration of the bacterium to the host (chemotactic activity), followed by the deployment of an iron sequestering system, resulting in eventual damage to fish by means of extracellular products, i.e., hemolysin and proteases (Haldar et al. 2010).

Pathogenesis of *Vibrio* consists of gaining access to the host tissue, colonization, and invasion (Lee et al. 2008). During colonization, the composition of the organisms, such as outer membrane proteins (OMPs), plays an important role in their adhesion to host cells. OMP is a unique component of the gram-negative bacterial cell wall. OMPs are transmembrane proteins, accounting for about half of all the outer membrane. OMPs of gram-negative bacteria, which form channels for small hydrophilic molecules, are known as porins (Chakrabarti et al. 1996). However little information is known about the outer membrane protein of *V. paraha*emolyticus, so it is understandable that many studies have been focused on amplifying genes responsible for these porins, to obtain these recombinant properties and understand their role in pathogenicity (Ye et al. 2010).

Daniela Ceccarelli et al. (2013) reported that *Vibrio parahaemolyticus*, autochthonous to estuarine, marine, and coastal environments throughout the world, is the causative agent of food-borne gastroenteritis. More than 80 serotypes have been described worldwide based on the antigenic properties of the somatic (O) and capsular (K) antigens. Serovar O3:K6 emerged in India in 1996 and subsequently was isolated worldwide, leading to the conclusion that the first *V. parahaemolyticus* pandemic had taken place.

The most common fish disease caused by bacteria belonging to the genus *Vibrio* is vibrosis. Rajesh Kumar et al. (2007) reported that *Vibrio anguillarum* produces a 38-kDa major outer membrane porin protein (OMP) for biofilm formation and bile-protected activity caused by Vibriosis. They reported that the gene encoding the porin was used to construct a DNA vaccine. The evaluation in Asian seabass (*Lates calcarifer* Bloch) is a common species in the aquaculture industry of the Indian coast and a potential resource.

White spot syndrome virus (WSSV) causes severe economic loss in the shrimp culture industry worldwide. Sarathi et al. (2007) silenced the VP28 gene of the WSSV of shrimp by bacterially expressed dsRNA. This technology enabled them to produce a uniform quality of the VP28dsRNA, which was used to protect the shrimp against WSSV.

White tail disease (WTD) causes severe economic losses in prawn hatcheries and farms, and mortalities were observed at 100% within 2 or 3 days. *Macrobrachium rosenbergii* nodavirus (MrNV) and extra small virus (XSV) have been identified as pathogenic agents, which are 27 and 15 nm in diameter, respectively. Sudhakaran et al. (2008) cloned and sequenced the capsid protein of an Indian isolate of an extra small virus from *Macrobrachium rosenbergii* for gene silencing.

Rajesh Kumar et al. (2008) examines the potential efficacy of a DNA vaccine against *Vibrio anguillarum* through the oral route using chitosan nanoparticle encapsulation. The porin gene of *V. anguillarum* was used to construct a DNA vaccine using pcDNA 3.1, a eukaryotic expression vector, and the construct was named pVAOMP38. The chitosan nanoparticles were used to deliver the constructed plasmid.

## 2.5 Conclusion

Infectious disease in aquatic organisms leads to a drastic change in the biome, thereby causing a potential threat to mankind who are dependent on it. Also, it leads to a severe and catastrophic blow to the economy of a nation. Focusing on the growing human population and the need for alternative resources (water-based cultivation) for survival, several researches focus on the development of various interventions that cure infection among aquatic organisms. On a cost-based analysis of various therapeutic measures, vaccines have several advantages and have proven to be cost-effective.

From various experimental trials, it is evident that vaccines play a significant role in boosting immunity and providing an immune response against many fish pathogens. Hence, this study has focused on the area of vaccination towards preventing and eradicating various disease pathogens that affect aquatic life. Also, attention is drawn towards the benefit of the implementation of vaccination over a large scale to cure several communicable diseases of aquatic fauna, which ultimately serves as an alternate food resource to the land-based food supply.

## References

Aasjord PM, Slinde E (1994) Fish vaccine: development, production and use of bacterial vaccines, with special reference to salmon. In: Martin AM (ed) Fisheries processing, biotechnological applicaiton. Chamen & Hall, London, pp 432–465

Anderson ED, Leong JC (2000) Development of DNA vaccines for slamonid fish. In: Lowrie DB, Whalen RG (eds) Methods in molecular medicine. DNA vaccines: methods and protocols. Humanana Press, Toltawa, pp 105–121

Aravena AR, Fuentes Y, Cartagena J, Brito T, Poggio V, Torre JL, Mendoza H, Nilo FG, Sandino AM, Spencer E (2015) Development of a nanoparticle-based oral vaccine for Atlantic salmon against ISAV using an alphavirus replicon as adjuvant. Fish Shellfish Immunol 45(1):1–10

Benmanzour A, de Kinkelin P (1997) Live fish vaccines; history and perspectives. In: Gudding R, Lillehaug A, Midtlyng PJ, Brown F (eds) Fish vaccinology developments in biological standardization, vol 90. Karger, Basel, pp 279–289

Boudinot P, Blanco M, de Kinkelin P, Benmasour A (1998) Combined DNA immunisation with the glycoprotein gene of viral hemorrhagic septicaemia virus and infectious hematopoietic necrosis virus induces double – specific protectice immunity and nonspecific response in rainbow trout. Virology 249(2):297–306

Ceccarelli D, Hasan NA, Huq A, Colwell RR (2013) Distribution of dyonamics of epidemic and pandemic vibrio parahaemolyticys virulence factors review article. Front Cell Infect Microbiol

Chakarabarti SR, Chaudhuri K, Sen K, Das J (1996) Porinsof vibrio cholera purification and characterization of OmpU. J Bacterial 178(2):524–530

Chen SN, Huang SL, G. H (1992) Studies on the epizootiology and pathogenicity of bacterial infections in cultured gaint tiger prawns. Penaeus monodon, in Taiwan. In: Diseases of cultured penaeid prawn in Asia and the United States. Hawaii Oceanic, Hawaii, pp 195–205

Corbeil S, Kurath G, LaPatra SE (2000) Fish DNA vaccine against infectious hematopoietic necrosis virus: efficacy of various routes of immunization. Fish Shellfish Immunol 10:711–723

Couch JA (1978) Diseases, parasites and toxic response of commercial penaeidprawns of the Gulf of Mexico and South Atlantic coasts of Atlantic coasts of North American. Fish Bull 76:1–44

Davis HL, McCluskie MJ (1999) DNA vaccines for viral diseases. Microbes Infect 1:7–21

De la Pena LD, Tamaki T, Momoyama K, Nakai T, Muroga K (1993) Characteristics of the causative bacterium for vibriosis in the kuruma prawn, Penaeusjaponocus. Aquaculture 115:1–12

Dehghani S, Akhlaghi M, Dehghani M (2012) Efficacy of formalin-killed, heat-killed and lipopolysaccharide vaccines against motile aeromonads infection in rainbow trout (*Oncorhynchus mykiss*) aquatic animal health unit, School of Veterinary Medicine, Shiraz University, shiraz, Iran. Glob Vet 9(4):409–415

Doll TA, Raman S, Dey R, Burkhard P (2013) Nanoscale assemblies and their biomedical applications. J R Soc Interf 10:20120740

Evelyn TPT (1997) A historical review of fish vaccinology. In: Gudding R, Lillehaug A, Midtlyng PJ, Brown F (eds) Fish vaccinology. Developments ion biological standardization, vol 90. Karger, Basel, pp 3–12

Frohlich E (2012) The role of surface charge in cellular uptake and cytotoxicity of medical nanoparticles. Int J Nanomedicine 7:5577–5559

Gomez Chiarri M, Livingston SK, Muro Cacho C, Sanders S, Levine RP (1996) Introduction of foreign genes into the tissue of live fish by direct injection and particle bombardment. Dis Aquat Org 27(1):5–12

Gregory AE, Titball R, Williamson D (2013) Vaccine delivery using nanoparticles. Front Cell Infect Mi 3:3389

Haldar S, Maharajan A, Chatterjee S, Hunter SA, Chowdhury N et al (2010) Identification of Vibrio harveyi as a causative bacterium for a tail rot disease of sea bream Sparusaurata, from research hatchery im Malta. Microbiol Res 165:639–648

Halperin S, Shefiele DE, Mcdonald NE, FRCPC (1992) Acellular versus whole-cell pertussis vaccines. Can J Infect Dis 3(2):57–58

Hanfman DT (1993) Aquaculture: January 1990 - July 1993. Beltsville, MD: National Agricultural Library

Heppell J, Davis HL (2000) Application of DNA vaccine technology to aquaculture. Adv Drug Deliv Rev 43:29–43

Jayasree L, Janakiram P, Madhavi R (2006) Characterization of Vibrio spp. associated with diseased shrimp from culture ponds of Andhra Pradesh (India). J World Aquacult Soc 37(4):523–532
Jiravanichpaisal P, Miyazaki T, Limsuwan C (1994) Histopathology, biochemistry and pathogenicity of *vibrio harveyi* infecting black tiger prawn, *Penaeus monodon*. J Aquat Anim Health 6:27–35
Jones SRM (2001) Plasmids in DNA vaccination. In: Schleet M (ed) Plasmids for therapy and vaccination. Wiley, Weinheim, pp 169–191
Kim YB, Okuda J, Matsumoto C, Takahashi N, Hashimoto S, Nishibuchi M (1999) Identification of vibrio parhaemolyticus strains at the species level by PCR targeted to the toxR gene. J Clin Microbiol 37(4):1173–1177
Kuzyk MA, Burian JM, Machander D (2001) An efficacious recombinant subunit vaccine against the salmonid rickettsial pathogen Piscirickettsia slamonis. Vaccine 19:2337–2344
Lai W, Hu Z, Fang Q (2013) The concerns on biosafety of nanomaterials. JSM Nanotech Nanomed 1:1009
Lavilla-Pitago CR, de la Pena LD (1998) Bacterial disease in prawn (*Penaeus monodon*) culture in Philippines. Fish Pathol 33:405–411
Lavilla-Pitago CR, Leano EM, Paner MG (1998) Mortalities of pond-cultured juvenile prawn, *Penaeus monodon,* associated with dominance of luminescent Vibrios, in the rearing environment. Aquaculture 164:337–349
Lee JK, Jung DW, Eom SY, Oh SW, Kim YJ, Kwak HS et al (2008) Occurrence of Vibrio parahaemolyticus in oysters from Korean retail outlets. Food Control 19:990–994. https://doi.org/10.1016/j.foodcont.2007.10.006
Leong JC, Anderson E, Bootland LM, Chiou PW, Johnson M, Kim C, Mourich D, Trobridge G (1997) Fish vaccine antigens produced or delivered by recombinant DNA technologies. Dev Biol Stand 90:267–277
Li L, Lin SL, Deng L, Liu ZG (2013) Potential use of chitosan nanoparticles for oral delivery of DNA vaccine in black sea bream Acanthopagrus schlegelii Bleeker to protect from Vibrio parahaemolyticus. J Fish Dis 36(12):987–995
Liang Z, Arjun S, Nani W, Chun XZ, Neena M, Chengzhong Y, Anton PJ, Middelber (2014) Nanoparticle vaccines. Vaccine 32:327–337
Lightner DV (1983) Disease of cultured penaeid prawn. In: Mcvey (ed) CRC hand book of mariculture, Crustcean aquaculture, vol 1. CRC Press, Boca Raton, pp 289–320
Lightner DV (1985) A review of the diseases of culture penaeid prawns and prawns with emphasis on recent discoveries and developments, pp 79–103
Lightner DV (1988) Vibrio disease of penaeidprawn. In: CJ, pp 42–47
Lightner DV (1996) Handbook of prawn pathology and diseases of cultured penaeidprawn. World Aquaculture Society, Baton Rouge
Lopez-Doriga MV, Smail DA, Smith RJ (2001) Isolaiton of Salmon pancreas disease virus (SPDV) in cell culture and its ability to protect against infection by the wild type agent. Fish Shellfish Immunol 11:505–522
Lorenzen E, Einer-Jensen K, Martinussen T, LaPatra SE, Lorenzen N (2000) DNA vaccination of rainbow trout against viral hemorrhagic septicemia virus: a dose-response and time-course study. J Aquat Anim Health 12(3):167–180
Lorenzen N, Lorenzen E, Einer-Jensen K, LaPatra SE (2002) DNA vaccines as a tool for analyzing the protective immune response against rhabdoviruses in rainbow trout. Fish Shellfish Immunol 12:439–453
Lorenzen E, Lorenzen N, Einer-Jenson K, Brudeseth B, Evenzen Ø (2005) Time course study of in situ expression of antigens following DNA vaccination against VHS in rainbow trout (Oncorhynchus mykiss, valbaum) fry. Fish Shellfish Immunol 19(1):27–41
Maiti B, Dubey S, Munang'andu HM, Karunasagar I, Karunasagar I, Evensen Ø (2020) Application of outer membrane protein-based vaccines against major bacterial fish pathogens in India. Front Immunol 11:1362. https://doi.org/10.3389/fimmu.2020.01362

Mamo T, Poland GA (2012) Nanovaccinology: the next generation of vaccines meets 21st century materials science and engineering. Vaccine 30:6609–6611

McLauchlan PE, Collet B, Ingerslev E, Secombes CJ, Lorenzen N, Ellis AE (2003) DNA vaccination against viral haemorrhagic septicaemia (VHS) in rainbow trout: size, dose, route of injection and duration of protection – early protection correlates with Mx expression. Fish Shellfish Immunol 15(1):39–50

Mialhe E, Bachere E, Boulo V, Cadoret JP, Rousseau C, CedeZo V, Saraiva E, Carrera L, Calderon J, Colwell RR (1995) Future of biotechnology-based control of disease in marine invertebrates. Mol Mar Biol Biotechnol 4:275–283

Mohamed S, Mona S, Magdy EM, Mansour EM (2016) Recent progress in applications of nanoparticles in fish medicine: a review. Nanomedicine 12:701–710

Mohney L, Bell A, Lightner DV (1994) An epizootic of vibriosis in Equadorian pond-reared *Penaeus vannamei* Boone (Crustacea:Decapoda). J World Aquac Soc 25:116–125

Munn CB (1994) The use of recombinant DNA technology in the development of fish vaccines. Fish Shellfish Immunol 4:459–473

Newman SG (1993) Bacterial vaccines for fish. Annu Rev Fish Dis 3:145–185

Nirmal M, Mariusz S, Istvan T (2014) Oral delivery of nanoparticle-based vaccines. Expert Rev Vaccines. https://doi.org/10.1586/14760584.2014.936852

Overstreet RM (1978) Marine maladies. Worms, germs and germs and other symbionts from the Norhtern Gulf of Mexico. Mississippi Alabama Sea Grant Consortium, Ocean Springs, Mississippi, USA

Pankhurst QA, Connolly J, Jones SK, Dobson J (2003) Application of magnetic nanoparticles in biomedicine. J Phys D Appl Phys 36:167–181

Pasnik DJ, Smith SA (2005) Immunogenic and protective effects of a DNA vaccine for *Mycobacterium marinum* in fish. Vet Immunol Immunopathol 103(3–4):195–206

Pridgeon JW, Klesius PH (2010) Fish vaccines in aquaculture–proactive treatment protects salmon, catfish, other fish. Global Aquaculture Advocate

Purcell MK, Kurath G, Garver KA, Herwig RP, Winton JR (2004) Quantitative expression profiling of immune response genes in rainbow trout following infectious haematopoietic necrosis virus (IHNV) infection or DNA vaccination. Fish Shellfish Immunol 17(5):447–462

Rajesh Kumar S, Parameswaran V, Ishaq Ahmad VP, Syed Mustaq S, Sahul Hameed AS (2007) Protective efficiency of DNA vaccination in Asian sea bass (Lates calcarifer) against vibrio anguillarum. Fish Shellfish Immunol 23:316–326

Rajesh Kumar S, Ishaq Ahmed VP, Parameswaran V, Sudhakaran R, Sarath Babu V, Hameed ASS (2008) Potential use of chitosan nanoparticles for oral delivery of DNA vaccine in Asian sea bass (Latescalcarifer) to protect from Vibrio (Listonella)anguillarum. Fish Shellfish Immunol 25:47–56

Ronen A, Perelberg A, Abramowitz J (2003) Efficient vaccine against the virus causing a lethal disease in cultured Cyprinuz Carpio. Vaccine 21:4677–4684

Ruangpan L, Kitao J (1992) Minimal inhibitory concentrations of 19 chemotherapeutants against Vibrio bacteria of prawn *Penaeus monodon*. In: Shariff M et al (eds) Disease of Asian aquaculture, pp 135–142

Sarathi M, Simon MC, Ishaq Ahmed VP, Rajesh Kumar S, Sahul Hameed AS (2007) Silencing VP28 gene of white spot syndrome virus of shrimp by bacterially expressed dsRNA

Schnick RA, Alderman DJ, Armstrong R, Le Gouvello R, Ishihara S, Lacierda EC, Percival S, Roth M (1997) Worldwide aquaculture drug and vaccine registration progress. Bull Eur Assoc Fish Pathol 17:251–260

Sindermann CJ (1990) Principal diseases of marine fish and shell fish, vol 2. Gulf Professional Publishing

Sommerset I, Skern R, Biering E, Bleie H, Fiksdal IU, Grove S, Nerland AH (2005) Protection against Atlantic halibut nodavirus in turbot is induced by recombinant capsid protein vaccination but not following DNA vaccination. Fish Shellfish Immunol 18(1):13–29

Sudhakaran R, Syed Musthaq S, Rajesh Kumar S, Sarathi M, Sahul Hameed AS (2008) Cloning and sequencing of capsid protein of Indian isolate of extra small virus from *Macrobrachium rosenbergii*. Virus Res 131:283–287

Tissot AC, Maurer P, Nussberger J, Sabat R, Pfister T, Ignatenko S (2008) Effect of immunisation against angiotension II with CYT006-AngQb on ambulatory blood pressure: a double-blind, randomized, placebo-controlled phase IIa study. Lancet 371:821–827

Vesely T, Pokorova D, Einer-Jensen K, Lorenzen N (2004) DNA vaccination of small carp (*Cyprinus carpio*) against SVCV: the protective effect depends on temperature. In: Book of Abstracts, 6th international symposium on fish immunology. 26–29 May, Turku, Finland, p 43

Vinay NT, Choudhury TG, Paria A, Gupta SK, Sarkar B (2016) Nanovaccines: a possible solution for mass vaccination in aquaculture. World Aquaculture. https://www.researchgate.net/publication/313846676_Nanovaccines_a_Possible_Solution_for_Mass_Vaccination_in_Aquaculture/citations

Vinay TN, Bhat S, Choudhury TG, Paria A, Jung MH, Kallappa GS, Jung SJ (2017) Recent advances in application of nanoparticles in fish vaccine delivery. Rev Fish Sci Aquac 26(1):29–41

Yang J, Wu X, Zc S (1992) Observations on black spot on shell disease of cultivated penaeidprawn by SEM. Donghai Mar Sci 11:34–39

Ye Y, Xiuli W, Sheping G, Xuemei Q (2010) Gene cloning and prokaryotic expression of recombinant outer membrane protein from *vibrio parahaemolyticus*. Chin J Oceanol Limnol 29(5):952–957

Yildirimer L, Thanh NT, Loizidou M, Seifalin AM (2011) Toxicological considerations of clinically applicable nanoparticles. NanoToday 6:585–607

You C, Han C, Wang X, Zheng Y, Li Q, Hu X (2012) The process of silver nanoparticles in antimicrobial mechanism, clinical application and cytotoxicity. Mol Biol Rep 39:9193–9920

Zaman M, Good MF, Toth I (2013) Nanovaccines and their mode of action. Methods 60:226–231

Zhang W, Zhu C, Xiao F, Liu X, Xie A, Chen F, Dong P, Lin P, Zheng C, Zhang H, Gong H, Wu Y (2021) pH-controlled release of antigens using mesoporous silica nanoparticles delivery system for developing a fish oral vaccine. Front Immunol 12:644396. https://doi.org/10.3389/fimmu.2021.644396

Zhao L, Seth A, Wibowo N, Zhao CX, Mitter N, Yu C, Middelberg APJ (2014) Nanoparticle vaccines. Vaccine 32:327–337

Zhu B, Zhang C, Zhao Z, Wang GX (2020) Targeted delivery of monosylated nanoparticles improve prophylactic efficacy of immersion vaccine against fish viral disease. Vaccine 8(1):87. https://doi.org/10.3390/vaccines8010087

# Chapter 3
# Nanovaccine

**Biswajit Maiti, Mave Harshitha, Somanath Disha, Anjana Kaveri Badekila, Sudarshan Kini, and Praveen Rai**

## 3.1 Introduction

Disease prevention by vaccination is one of the landmarks of modern medicine. Vaccination is a method of producing an active immune response in an organism against a targeted pathogen. It is one of the most influential and sustainable methods to treat bacterial and viral diseases in humans as well as in veterinary medicine. Aquaculture produces ~80 million tonnes of aquatic animals with a first-sale value of $232 billion. It represents the fastest-growing animal production sector in the world, making aquaculture the fastest primary food-producing sector globally. However, the major hindrance faced in the industry is the incidence of diseases, which amounts to more than $10 billion in losses annually on a global scale (http://www.fao.org/3/i9540en/I9540EN.pdf). The disease outbreaks may be due to various pathogenic organisms like viruses, bacteria, fungi, and parasites, which are generally tackled with the help of chemical agents, antibiotics, and vaccines. Lately, nanotechnology is being applied in many fields of science, including vaccine development. Nanotechnology refers to the study and use of structures between 1 and 100 nm in size. The technology has been applied and is undergoing clinical trials to deliver many therapeutics such as antibiotics and prophylactic treatments. The increasing significance of nanotechnology-based drug delivery systems has been an essential aspect of researching and developing various novel vaccine formulations.

B. Maiti (✉) · M. Harshitha · S. Disha · P. Rai
NITTE (Deemed to be University), Nitte University Centre for Science Education and Research (NUCSER), Division of Infectious Diseases, Mangaluru, Karnataka, India
e-mail: maiti.b@nitte.edu.in

A. K. Badekila · S. Kini
NITTE (Deemed to be University), Nitte University Centre for Science Education and Research (NUCSER), Division of Nanobiotechnology, Mangaluru, Karnataka, India

 
V. K. Arivarasan et al. (eds.), *Nanotechnological Approaches to the Advancement of Innovations in Aquaculture*, Nanotechnology in the Life Sciences, https://doi.org/10.1007/978-3-031-15519-2_3

Nanovaccines are a new generation of vaccines consisting of nanoparticles (NPs) and a pathogen-specific antigen that elicit a controlled immune response. Conventionally, in aquaculture, the antigens used in the vaccine formulations are peptides, nucleic acids, toxoids, and other biomolecules. But due to the selective permeability of the cell membranes, most of these biological macromolecules cannot enter the cells to activate an effective immune response. As they have a large surface-to-volume ratio, nanoparticles can overcome this limitation and enhance circulation time, promote bioaccumulation in lymphoid organs, and efficiently target immune cells (Bharadwaj et al. 2020). The physio-chemical properties of the nano-carriers can be altered to give optimal antigen presentation, biodistribution, and cellular trafficking (Vartak and Sucheck 2016). In the past decade, there has been a high demand for nanoparticle-based vaccines due to increasing diseases in aquaculture globally. Nanovaccines are emerging as an improved and novel construct of vaccines.

## 3.2 Controlling the Disease Burden in Aquaculture Through Vaccinations

Aquaculture production peaked at 82.1 million tonnes in 2018, up by 3.2% from 2017 (http://www.fao.org/3/i9540en/I9540EN.pdf). In most countries around the world, the diseases in aquaculture were conventionally treated using antiviral drugs, antibiotics, antifungal agents, and other chemicals. This type of treatment can lead to various other undesirable effects like environmental pollution and multidrug resistance in bacteria. On the other hand, vaccines seem to be more effective and provide a long-term measure to treat diseases.

Vaccines may be defined as biological preparations that contribute to active immunity and long-term protection against a particular disease. The first vaccine in aquaculture was licensed in 1976 against enteric redmouth disease caused by *Yersinia ruckeri.* Currently, there are more than 20 commercially available licensed vaccines for aquaculture. These include whole killed vaccines, subunit vaccines, recombinant protein vaccines, DNA vaccines, and live attenuated vaccines. These vaccines are administered to fishes orally, by immersion, or by injection.

Live attenuated vaccines contain the pathogen in its weakened (attenuated) form. Earlier, the pathogens were passed through *in vitro* systems multiple times to attain random mutations (Adams 2019), which would weaken the pathogen. But in recent times, this attenuation is obtained through the genetic modification of the organism. Generally, a single dose of this vaccine is sufficient to elicit a robust immune response. Live attenuated vaccines are more beneficial in the case of diseases caused by intracellular pathogens. However, safety and reversion into the virulent form are always major concerns.

In the case of salmonids, vaccines have been used for almost 30 years. Most of them are oil-adjuvant-based injectable vaccines. In the United States, there are currently two live attenuated vaccines available commercially for enteric septicaemia of catfish (caused by *Edwardsiella ictaluri*) and columnaris disease (caused by *Flavobacterium columnare*). There is another licensed vaccine against enteric septicaemia of catfish in Vietnam that is manufactured by PHARMAQ AS. Inactivated vaccines are available against photobacteriosis caused by *Photobacterium damselae* subspecies *piscicida* (by Merck). Inactivated vaccines are also available against vibriosis caused by *Vibrio anguillarum*, *Vibrio ordalii*, and *Vibrio salmonicida*. A virulent live culture is used as a vaccine against bacterial kidney disease caused by *Renibacterium salmoninarum*. An inactivated vaccine for Tenacibaculosis caused by *Tenacibaculum maritimum* in Spain (ICTHIOVAC® TM – HIPRA) is available.

Vaccines are also present for viral diseases such as koi herpesvirus (CyHV-3) – there is a subunit vaccine (peptide-VP2) against infectious pancreatic necrosis in Norway (Merck) – viral haemorrhagic septicaemia. The first DNA vaccine to be licensed for use in aquaculture was Apex-IHN (Elanco) against infectious haematopoietic necrosis virus (IHNV) in Atlantic salmon (Salonius et al. 2007). There are other vaccines against salmon pancreatic disease (Merck), such as a recombinant vaccine against infectious salmon anaemia caused by salmon *isavirus* (Centrovet) in Chile.

Vaccines in aquaculture are mainly administered by the immersion method, oral administration, and intraperitoneal injections. Most of the vaccines available currently are of the whole-cell killed pathogen type, which is injected intraperitoneally. However, this type of vaccine might not be feasible in all countries; as a result, most of them settle with antibiotics. Therefore, there is a significant need for more knowledge and research in this field. The more feasible oral administration of drugs essentially requires protection as the antigenic components may be subjected to gastric digestion or digestion before absorption (Aklakur et al. 2016). Vaccines activate an immune response against the respective diseases but can have restrictions such as compromised effectiveness, reduced immunogenic responses, poor constancy, and the need for booster doses. These limitations in conventional vaccines have surfaced a need for an improved form of vaccines. With the recent advancements in nanotechnology, science has paved the way to explore the blend of nanotechnology and immunotherapy through nanovaccines.

## 3.3 Choice of Antigens as Nanovaccines

Nanovaccines mainly consist of two components, i.e. a synthetic or natural nanomaterial that functions as a carrier/adjuvant and an antigen. The antigens can be of various subunits of the pathogen types such as peptides, proteins, polysaccharides, capsules, and toxins, but ultimately, all of them have the collective objective of eliciting an immune response against the targeted pathogen. There are inactivated or killed forms of the pathogens which are used to activate an immune response. Formalin-killed cells are primarily used in aquaculture. Although these cells can activate both humoral and cell-mediated immunity, they sometimes do not attain an optimal level of protection (Collins et al. 2019). Subunit vaccines do not contain any whole bacteria or viruses; instead, they have one or more specific antigens of the pathogen that function as 'flags' for the immune system. Subunit vaccines are mainly designed through reverse vaccinology, which involves identifying a suitable vaccine candidate through *in silico* analysis.

Outer membrane proteins (OMPs) present on the membrane of gram-negative bacteria have become an important topic of research. They are present in the outermost region of the bacteria and are the first to come into contact with the host and help in adhesion, invasion, and contributing to pathogenesis. They are conserved among serotypes and are highly immunogenic in nature; hence, they would serve as an efficient vaccine candidate (Maiti et al. 2012, 2020).

DNA molecules have tremendous potential as vaccine candidates. DNA vaccines elicit both humoral and cell-mediated immune responses. Still, they sometimes lose their integrity before they reach the target cell due to degradation by endogenous nucleases, and also due to the net negative charge present on the cell surface, the DNA molecules can be repelled away from the cell membrane (Bhavsar and Amiji 2007). Hence, nanoparticles can ensure this integrity for effective delivery of the polynucleotide to the target site, followed by cellular internalization and processing. It is more beneficial to deliver DNA vaccines with the help of polymer-based nanoparticles (Bhavsar and Amiji 2007). Similarly, RNA-based vaccines are also safe and effective. As the name suggests, virus-like particles are particles that resemble the external structure of a virus but do not contain any genetic material; hence, they cannot replicate or mutate. They have the ability to self-assemble into the full tertiary structure that mimics the virus (Jeong et al. 2020). The antigen-presenting cells (APCs) recognize the epitopes on the particle to generate an immune response. They will require booster doses as the particles cannot replicate.

## 3.4 Nanoformulation of Vaccines and Delivery Systems

A revolutionization in the food and aquaculture sectors due to the huge potential of nanotechnology has provided innovative tools for disease prevention. Despite the tremendous application of traditional vaccines, lack of control over vaccine release, intrinsic instability, indiscriminate distribution, and a need for multiple administrations in aquaculture are various concerns governing the health sector today. To overcome these limitations, a nanoparticle-based drug delivery system provides a tool for targeting the desired site and enhancing the immune response of the host against the invading pathogens. Generally, the generation of vaccines with nanoparticles as adjuvants is formulated by optimizing the size, loading capacity, and surface charge of the nanoparticles. Further, an optimized dosage of nanovaccine, in turn, stimulates the immune system, thereby triggering the adaptive immune response during pathogen invasion. Previous findings suggest that the application of nanovaccines compared to traditional vaccines showed maximal antibody production in animal models (Gheibi Hayat and Darroudi 2019). Hence, novel nanovaccines are enormously necessary for sustaining aquatic health, preserving endangered species, and inhibiting the passage of pathogens in the food chain (food-borne illness).

Nanoparticles have been used as drug delivery vehicles for fish immunization studies. The use of polymeric nanoparticles has the advantage of allowing bioactive molecules to be encapsulated and protected from hydrolytic and enzymatic degradation. Due to their rapid escape from the degradative endo-lysosome, nanoparticles containing plasmid DNA provide sustained release. The use of a nanoparticle vaccine delivery system can also help to improve the immune response to synthetic peptide vaccines (Salvador et al. 2011). Biologics, polymers, carbon-based materials, silicon-based materials, and metals are commonly used in alternatives for drug delivery (nanoscale). Protein and gene delivery are being investigated using biodegradable polymer nanoparticles made of polylactic acid (PLA), polyglycolic acid (PGA), or polylactic-co-glycolic acid (PLGA). Some polymers under investigation for nanoscale drug carriers consist of poly (3-hydroxybutanoic acid), polyglycolic acid, poly (ethylene glycol), poly (ethylene oxide), and copolymers such as PLA-PEG. Solid nanoparticles have been engineered using a variety of materials, both with and without surface functionality. The aliphatic polyesters, specifically the hydrophobic PLA, the hydrophilic PGA, and their copolymer PLGA, can make up the majority of polymers. Polymeric NPs are made up of a polymer that has gained much deliberation due to advancements in polymer science and technology advancements. Biocompatibility and biodegradability are critical characteristics for tissue engineering, drug and gene delivery, and novel vaccine strategies. Antigen/drug delivery using polymeric nanoparticles has many benefits over conventional delivery methods, including the ability to target drug delivery to a particular site, such as an intracellular infection (Abdelghanyet al. 2012), reducing systemic toxicity, and facilitating the sustained release of a drug.

## 3.5 Different Types of Nanoparticles and Their Potential Applications for the Delivery of Vaccines

Today, different nanostructures are applied as delivery vehicles for vaccine efficacy and to explore newer avenues in vaccine administration in aquaculture (Fig. 3.1). The most applied nanoparticles as adjuvants include polymeric nanoparticles like alginate, chitosan, PLGA, dendrimers, and liposomes. However, conjugating the vaccine molecules with appropriate nano-carriers is essential to improve vaccine's characteristics and its delivery potential. Conjugation of vaccine molecules with a suitable nano-carrier is generally performed by surface conjugation, encapsulation, and surface adsorption. The presence of a charge or hydrophobic interaction between nanoparticles and antigen molecules provides a medium for antigen adsorption on the surface of a nano-carrier. Usually, due to the non-covalent nature of their interaction, a sudden dissociation of antigens from the nano-carriers owing to the effect of external factors like pH, ionic strength, and temperature makes surface adsorption less popular. Other methods like encapsulation by embedding the antigen of interest and surface conjugation are universally applied in nanovaccine preparations

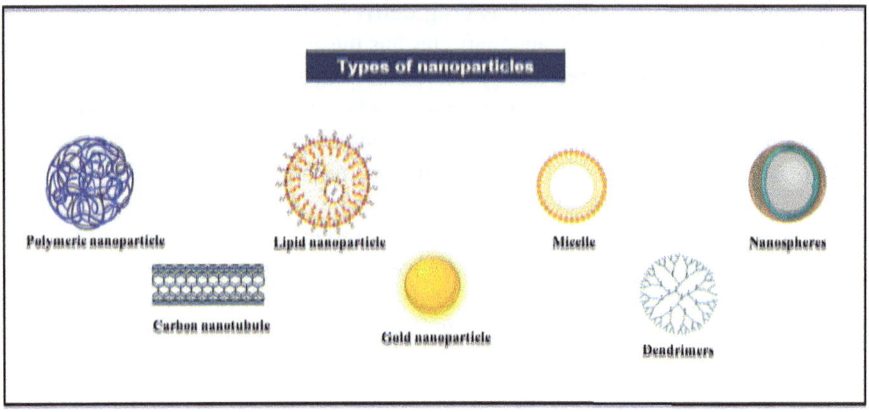


**Fig. 3.1** Different types of nanoparticles. Created with BioRender.com

due to strong intermolecular bonding, surface interaction, and the ability to release the antigen after partial or complete degradation of the nano-carrier (Pati et al. 2018). The nanovaccines used against various fish pathogens are summarized in Table 3.1.

Once the desired vaccine is designed and formulated, effective administration of the immunostimulant to control the infectious diseases in fish is necessary. Techniques may be many; however, there are three popular methods for vaccine administration in fish: oral, immersion, and injectable administration. Nowadays, vaccination by injection is the most reliable and effective method in fish immunization when compared to immersion and oral means, owing to easier administration, stress-free handling, and dose determination ability. Nanoparticles as delivery systems provide a medium for administering vaccines in a controlled manner to achieve the desired therapeutic effect. This is because cell or tissue targeted delivery, bioavailability, enhanced solubilization of hydrophobic drugs, controlled release, and therapeutic agent dissemination from degradation are protected (Ji et al. 2015). In this chapter, a summarization of different nano delivery systems for fish vaccination is further described.

### *3.5.1 Polylactic-Co-Glycolic Acid*

A biodegradable polymer, PLGA, is a copolymer of lactic acid and glycolic acid. The characteristics like degradation rate, the strength of the nano-carrier, and loading capacity are altered by adjusting the monomeric ratio (Badekila et al. 2021). The biological compatibility and biodegradable nature of PLGA are helpful for human use and are approved by the Food and Drug Administration (FDA, USA) and the European Medicines Agency. This is due to the hydrolysis of PLGA into glycolic acid and lactic acid monomers, which are easily excreted (citric acid cycle) from the

**Table 3.1** Different types of nanovaccines against fish pathogens

| Antigen | Pathogen | Nanoparticle | Species | Delivery Route | RPS (%) | References |
|---|---|---|---|---|---|---|
| *Vibrio harveyi* | *Vibrio harveyi* | Liposome | *E. bruneus* | IP | 75 | Harikrishnan et al. (2012a, b, c) |
| NKC03,IKC03 | KHV | Liposome | *C. carpio* | Oral | 74.3 | Yasumoto et al. (2006) |
| *Aeromonas salmonicida* (T1031) | *Aeromonas salmonicida* | Liposome | *C. carpio* | Oral | 83.3 | Irie et al. (2005) |
| pcDNA-vp7 | GCRV | Carbon nanotubes | *Ctenopharyngon idellus* | IM | 72.5 | Zhu et al. (2015) |
| pEGFP-vp5 | GCRV | Carbon nanotubes | *C. idellus* | IM | 56.7 | Wang et al. (2015) |
| rVP7 | GCRV | Carbon nanotubes | *C. idellus* | Bath | 37.7 | Zhu et al. (2014) |
| MOMP | *A. hydrophila* | ISCOMs | *Anguillia anguillia* | IP | 80 | Dong et al. (2005) |
| S-layer protein | *A. hydrophila* | Calcium phosphate | *Labeo rohita* | IP | 100 | Behera and Swain (2011) |
| pVAOMP38 | *V. anguillarum* (Listonella) | Chitosan | *L. calcarifer* | Oral | 46 | Rajesh Kumar et al. (2008) |
| pVAOMP-DNA | *V. anguillarum* | Chitosan/ TPP | *L. calcarifer* | Oral | ND | Vimal et al. (2012) |
| pFNCPE42 | Nodavirus | Chitosan/ TPP | *L. calcarifer* | Oral | 60 | Vimal et al. (2014) |
| OmpK | *V. parahaemolyticus* | Chitosan/ TPP | *A. schlegelii* | Oral | 60 | Vimal et al. (2014) |
| rgpG | VHSV | Chitosan/ TPP | *D. rerio* | IP | 70 | Kavaliauskis et al. (2016) |
| pEGFP-N2-TRBIV-MCP (pDNA) | TRBIV | Chitosan/ TPP | *Scophthalmus maximus* | Oral | ND | Zheng et al. (2016) |
| ISAV (V) | ISAV | Chitosan/ TPP | *Salmo salar* | Oral | 40.4 | Zheng et al. (2016) |
| TNP-LPH | TNP-LPH | PLGA | *S. salar* | IP | ND | Fredriksen et al. (2011) |
| pEGFP-N2-MCP | LCDV | PLGA | *P. olivaceus* | Oral | ND | Tian and Yu (2011) |
| HGG | LCDV | PLGA and PLA | *S. salar* | IP | ND | Fredriksen and Grip (2011) |
| pCDNA-G | IHNV | PLGA | *O. mykiss* | Oral | 22 | Adomako et al. (2012) |
| TA,PT | IPNV | PLGA | *S. salar* | IP | 16.7 | Munang'andu et al. (2012) |

(continued)

**Table 3.1** (continued)

| Antigen | Pathogen | Nanoparticle | Species | Delivery Route | RPS (%) | References |
|---|---|---|---|---|---|---|
| Omp | *A. hydrophila* | PLGA, PLA | *L. rohita* | IP | 75 | Rauta and Nayak (2015) |
| SIP | *Streptococcus agalactiae* | PMMMA-PLGA | *Oreochromis niloticus* | Oral | 100 | Zhang et al. (2016) |
| rOmpW | *A. hydrophila* | PMMMA-PLGA | *L. rohita* | Oral | 79.9 | Dubey et al. (2016a, b) |
| rOmpA | *E. tarda* | Chitosan | *L. fimbriatus* | Oral | | Dubey et al. (2016a, b) |
| Irradiated trophont | *I. multifiliis* | Alginate | *O. mykiss* | Oral | | Heidarieh et al. (2015) |

*PMMMA* Poly [(methyl methacrylate)-co-(methylacrylate)-co-(methacrylic acid)], *PLA* Poly (lactic acid), *ISCOMs* Immunostimulating complexes, *PLGA* Poly (lactic-co-glycolic acid), *OCMCS* Oleoyl-carboxymethyl-chitosan, *bVHSV* Viral haemorrhagic septicaemia virus, *TRBIV* Turbot reddish body iridovirus, *GCRC* grass carp reovirus, *ISAV* Infectious salmon anaemia virus, *IPNV* Infectious pancreatic necrosis virus, *IHNV* Infectious hematopoietic necrosis virus, *LCDV* Lymphocystis disease virus, *KHV* Koi herpes virus, *IM* Intramuscular, *IP* Intraperitoneal, *dRPS* Relative percentage survival, *ND* Not determined

fish. It is one of the most commonly used biodegradable synthetic polymer nanocarriers, with a long history of biomedical application and a high safety profile (Semete et al. 2010). These particles are attractive for oral vaccination because they are easy to manufacture and relatively inexpensive. In the body, it undergoes non-enzymatic hydrolysis, yielding biodegradable metabolites such as lactic acid and glycolic acid. These are generally present in the body and participate in several biochemical and physiological pathways. As a result, there is very little systemic toxicity associated with PLGA use. PLGA and PLA can be made in various sizes and forms, and they can be used to encapsulate a wide range of molecules.

The size of the PLGA particles can be modified, and surface modifications can be added to vaccine formulations for oral, mucosal, and systemic administration. PLGA can easily be degraded in the body since, in the aqueous environment, its ester bond is hydrolyzed, allowing the easy release of antigens. The release kinetics of the system can be easily manipulated by varying the ratio of PLA/PLGA and is faster in acidic conditions. The particle size, surface alteration, and release profile of the PLGA nanoparticle all influence the immunogenicity of the entrapped antigen. PLA, unlike PGA, has a methyl group, making the copolymer more hydrophobic at higher PLA proportions. Similarly, the PGA/PLA ratio affects other particle physicochemical properties, including mechanical force, hydration power (swelling behaviour), gelation temperature, and charge.

PLGA is commonly prepared by the double emulsion method by dissolving it in an organic solvent like chloroform, dichloromethane, or ethyl acetate (McCall and Sirianni 2013). Hydrophobic compounds (antigens) are directly added to the organic phase, while hydrophilic compounds are emulsified with PLGA polymer solution prior to particle formation. Emulsification of the solution is performed by adding a surfactant or emulsifying agent like polyvinyl alcohol. The solid nanoparticles containing the antigen of interest are obtained by evaporating the solvent by continuous stirring or pressure reduction. However, the uptake mechanism of PLGA-loaded antigens is not well understood in aquatic culture (Ji et al. 2015). Recently, a study encapsulated an inactivated virus in PLGA nanoparticles and evaluated its efficacy against viral haemorrhagic septicaemia virus (VHSV) in the marine fish, *Paralichthys olivaceus* (12 g) by oral/immersion route of administration. A relative percent survival (RPS) greater than 60% was observed after 4 weeks in this study with an upregulation of immunity genes like IgM, IgT, pIgR, MHC-I, MHC-II, IFN-γ, and Caspase 3 (Kole et al. 2019). Another study investigated a PLGA encapsulated vaccine for innate and adaptive immune responses in kelp grouper, *Epinephelus bruneus* (31 ± 2 g), against an opportunistic pathogen in marine fishes, *Uronema marinum*, at different time intervals of 1–4 weeks. The cumulative mortality was less than 20% in PLGA-based vaccinated groups during scuticociliatosis disease. The acquired protection of the marine fishes against the protozoans was significantly due to the PLGA-based vaccination that comparatively enhanced respiratory burst activity, complement activity, and α-2-microglobulin (Harikrishnan et al. 2012a, b, c). Till today, high survival and serum antibody content suggested both direct and indirect involvement of PLGA vaccine in stimulating immune protective factors. A further mechanism of long-term protection needs to be evaluated for the future enhancement of aquatic organisms.

PLGA particles may also serve as adjuvants (Katare and Panda 2006), and the FDA has approved their use in human and veterinary medicine. Multiple antigens can be released simultaneously with PLGA particles, and antigens can be transported to intracellular compartments. To activate APCs, they may be engineered to be the same size as. Furthermore, biological degradation can take months to years depending on their characteristics (Prokop and Davidson 2008). Many researchers have studied the feasibility of using PLGA as a drug carrier. In a study, PLGA nanoparticles were filled with the anti-mycobacterial agent rifampicin and then injected into zebrafish embryos. Since the zebrafish embryos are transparent, it showed that the treatment had a significant impact on *Mycobacterium marium* infection. The rifampicin-PLGA nanoparticle showed an increased therapeutic effect against *M. marium* and higher embryo survival compared to rifampicin alone (Danhier et al. 2012).

### 3.5.2 Polylactic Acid

PLA is a biocompatible and biodegradable polymer. It is non-toxic and can be metabolized into monomeric units of lactic acid in the body. Since 1980, PLA polymer has been extensively studied in surgical implants, sutures, and drug delivery.

### 3.5.3 Alginate

Alginate is a natural polymer derived from brown seed weed/algae that comprises copolymers containing D-mannuronic acid and -L-guluronic acid found in brown algae cell walls. It is biodegradable, biocompatible, non-toxic, acid-resistant, mucoadhesive, and well-suited for oral vaccine administration (Rivas-Aravena et al. 2013). Alginate is commonly used in antigen encapsulation for a variety of reasons: it has low toxicity and mucoadhesiveness, it allows interaction of the alginate particle with the epithelial mucus walls; it is acid and protease-resistant, and it is inexpensive. The size, antigenic composition, production strategy, alginate selection, and antigen concentration influence the alginate particle's characteristics. Alginate has been used in the delivery of fish vaccines in the form of microparticles. More often than nanoformulations, alginate nanoparticles were evaluated for oral vaccine delivery against *Ichthyophytirius multifiliis* in rainbow trout for booster vaccination.

The mechanical property of alginate polymer is dependent on the G units in its block formation. Alginate as cargo for vaccine delivery is attractive. It is stable at low pH during its release in the foregut and midgut of fish and can be applied in the oral administration of nanovaccines (Ji et al. 2015). A study combined alginate and chitosan by ionic gelation for oral vaccination of *Oncorhynchus mykiss* (10 g) against pathogenic bacteria like *Lactococcus garvieae* and *Streptococcus iniae*. The challenge test results showed 76% survival (against *S. iniae*) and 66% survival (against *L. garvieae*) for the polymer-based vaccine after 10 days of the challenge. Thus, indicating the effectiveness of the oral bivalent *Streptococcus-Lactococcus* vaccine (Halimi et al. 2019). Many studies have applied alginate in microform; however, studies of alginate as a nanovaccine are still in their nascent stages. Hence, further improvement in the application of alginate-based vaccines in nanoparticle form needs to be performed.

### 3.5.4 Chitosan

Chitosan is a linear polysaccharide composed of (1–4)-linked D-glucosamine and N-acetyl-D-glucosamine, which is present in the exoskeletons of crustaceans, insects, and some microorganisms. Chitosan nanoparticles are used for drug delivery and have excellent properties. They are made of a biocompatible, non-toxic, and biodegradable polymer that is easily excreted by the kidneys. Because of their mucoadhesive properties, they can be modified for slow and sustainable drug release. In reality, chitosan activates immune cells such as macrophages, natural killer cells, APCs, and T lymphocytes by stimulating the production of cytokines (Foged et al. 2005). Purity, molecular weight, degree of deacetylation, quality, and viscosity all affect the characteristics of chitosan particles. Fish vaccines have been developed using chitosan nanoparticles, such as the inactivated virus vaccine against infectious salmon anaemia virus (ISAV), which includes the DNA coding for ISAV replicase as an adjuvant. The outer membrane protein K-encoding gene of *Vibrio parahaemolyticus* was loaded onto chitosan nanoparticles to create an oral DNA vaccine. In the black sea bream *Acanthopagrus schlegelii*, this recombinant nanovaccine elicited a defensive immune response against *V. parahaemolyticus* (Li et al. 2013). Recently, a reddish body iridovirus oral DNA vaccine based on chitosan nanoparticles was created (Zheng et al. 2016). Chitosan and chitosan/triopoly phosphate nanoparticles were used to produce an oral DNA vaccine against *V. anguillarum* in Asian sea bass, *Lates calcarifer*. The nanovaccine conferred only moderate protection against the pathogen (Vinay et al. 2016). Similarly, in Asian sea bass (*L. calcarifer*), chitosan nanoparticles were tested for their ability to deliver plasmid DNA encoding *V. anguillarum* OMPs (Poobalane et al. 2010). Using immunohistochemistry, Omp38 was subsequently found in the liver, kidney, spleen, and intestine. Fish were partly safe from homologous challenge 21 days after vaccination, with an RPS of 46%. In a study on rainbow trout *(O. mykiss)*, vitamin C was found to be conjugated with chitosan nanoparticles (*O. mykiss*). Because of the potent synergism between chitosan and vitamin C, the vitamin was released up to 48 h after oral administration, and the fish's innate immune system was stimulated. It also has the potential to induce a robust adaptive immune response against the conjugated antigen, both cellular and humoral (Arca et al. 2009). The use of chitosan nanoparticles in fish vaccines has several advantages, including the ability to boost mucosal immunity through the oral route of vaccination. Several oral DNA vaccination studies in fish have shown that chitosan nanoparticles are more effective than other formulations against antigens derived from turbot reddish body iridovirus, nodavirus, *V. parahaemolyticus*, and *V. anguillarum*. Vimal et al. (2012) and Rivas-Aravena et al. (2015) also demonstrated the efficacy of chitosan nanoformulations against inactivated ISAV. A recent study found that chitosan was effective in the intraperitoneal administration of a vaccine against VHSV recombinant glycoprotein (rgpG) in zebrafish (Kavaliauskis et al. 2016). According to the findings, the use of chitosan nanoparticles improves vaccine-mediated protection against infection in fish. It also contributes to the immune response by modulating leukocyte trafficking. It is biocompatible, biodegradable, hydrophilic, and abundant in nature, making it one of the most appealing candidate nanoparticles.

### 3.5.5 Dendrimers

Dendrimers are symmetric nanoparticles constituting a central core, an inner shell, and an outer shell with treelike arms and branches. The synthesis of the dendrimer is generally performed by convergent (addition of monomers from the chain end) and divergent (synthesis starting from the core) methods followed by size and branch optimization (Klajnert and Bryszewska 2001). The presence of multiple functional groups in the dendrimer surface is advantageous in coupling biologically relevant molecules. With a need for efficient nanovaccines, dendrimers provide molecularly defined multivalent scaffolds to fabricate conjugates with antigens. The biocompatibility, predictable biodistribution, and ligand-receptor interacting characteristics are dependent on the size and surface charge of dendrimers (Heegaard et al. 2010). Hence, an optimized synthesis of dendrimers based on size and surface charge provides feasibility for efficient nanovaccine preparation in conjunction with delivery systems. Previously, few studies explored the suitability of amine–/amide-based dendrimers like polypropylene imine and polyamido amine (PAMAM) for antigen delivery to produce an immune response in the host against contagious viruses (Chahal et al. 2016). Currently, chlamydial infection in fish is emerging as a cause for concern in aquaculture industries (Stride et al. 2014). In this regard, the dendrimer-conjugated peptide vaccine is suitable for the clearance of the infection. The proposed carrier system in a previous study involved a PAMAM dendrimer to which a chlamydial peptide mimic, glycolipid antigen-peptide 4, was conjugated through an ester bond (Ganda et al. 2017). Even though nanoformulation is effective in controlling infectious bacterial and viral strains in mouse models, considerable research is not available in aquatic models.

### 3.5.6 Inorganic Nanoparticles

Inorganic nanoparticles have a measurable impact on modern material science research due to their possible technological importance, particularly in the field of bio-nanotechnology, and their unique physical properties including size-dependent magnetic, optical, electronic, and catalytic properties. Thus, spreading their possible applications in fluorescence labelling, magnetic resonance imaging, and stimulus-responsive drug delivery is crucial to the diagnosis and treatment of diseases (Kobayashi et al. 2014). It is a viable alternative to organic forms that could be used for vaccine delivery. Many inorganic nanoparticles have been explored for their application in vaccine delivery. Despite their non-biodegradable nature, inorganic nanoparticles have advantages like a unique rigid structure for controllable synthesis (Kalkanidis et al. 2006). The four most commonly used inorganic NPs are (1) noble metal, (2) magnetic, (3) fluorescence, and (4) multifunctional, e.g. luminescent magnetic.

### 3.5.7 Gold Nanoparticles

There is a great interest in investigating the antimicrobial effect of gold nanoparticles due to their low toxicity to eukaryotic cells. It can be used in vaccine delivery systems and can be easily manipulated into different sizes and shapes. Gold nanoparticles were produced by green synthesis and showed antibacterial activity against fish bacterial isolates.

### 3.5.8 Silver Nanoparticles

Silver nanoparticles are one of the most investigated nano-antibacterial agents in the research literature. It is synthesized using citrus (lemon) juice as a reducing agent. It has shown antibacterial activity against *Staphylococcus aureus* and *Edwardsiella tarda* and anti-cyanobacterial activity towards *Anabaena* and *Oscillatoria* species; there is very little published work on the antifungal and antiviral effect of silver nanoparticles in fish medicine.

### 3.5.9 Zinc Oxide Nanoparticles (ZnO-NPs)

Zinc oxide nanoparticles have drawn more attention because of their antibacterial and antifungal effects. In the field of fish medicine, ZnO-NPs can inhibit the growth of *Aeromonas hydrophila, E. tarda, Flavobacterium branchiophilum, Citrobacter* spp., *S. aureus*, *Vibrio* species, *Bacillus cereus*, and *Pseudomonas aeruginosa.* Ramamoorthy et al. investigated the antibacterial effects of ZnO-NPs against the pathogenic *Vibrio harveyi* and observed higher bactericidal effects of nanoparticles compared to bulk ZnO. In another interesting study, ZnO nanoparticles were synthesized biologically using *A. hydrophila*. These nanoparticles exhibited antibacterial activity against the same bacterium and other species like *P. aeruginosa*, *E. coli*, *Enterococcus faecalis*, *Aspergillus flavus*, and *Candida albicans.*

### 3.5.10 Titanium Dioxide Nanoparticles ($TiO_2$_NPs)

$TiO_2$-NPs, when doped with magnetic $Fe_3O_4$-NPs, had a bactericidal effect against *S. iniae*, *E. tarda*, and *P. damselae* after activation by light (Abdelghany et al. 2012). These particles can be used to disinfect water, as the fish pathogens bind with the nanoparticles, which can then be easily extracted from the water using a magnet (Abdelghany et al. 2012). However, Jovanovic et al. (2011) concluded that $TiO_2$-NPs influence the immune system of fish by decreasing the antimicrobial activity of fish neutrophils, rendering the fish more susceptible to infection and hence increasing mortality, particularly during disease outbreaks.

### 3.5.11 Nano-selenium

Nano-selenium has a substantial impact on the physiology of fish by improving the animal's physiological and immunological systems. Selenium supplementation protects cells from damage and is important for fish development, fertilization, and immunological function. By increasing lysozyme activity and red blood cell count, selenium supplementation boosts fish immunity. The nano form of selenium has the most favorable effect, as it is more effective than the bulky version. The nano form of selenium is a unique kind that draws greater interest than inorganic and organic forms due to its high bioavailability and lower toxicity (Khurana et al. 2019), whereas inorganic compounds are more harmful than organic compounds. The biological properties of selenium nanoparticles are dependent on their size; smaller particles have more activity. In comparison to other organic and inorganic oxidation states, nano-selenium (nano-Se) advantages from the capacity to utilize selenium at zero oxidation, which has low toxicity and excellent bioavailability. It's highly unstable, and it can readily revert to a dormant state. On the other hand, encapsulation with chitosan can help to stabilize it (Nasr-Eldahan et al. 2021).

### 3.5.12 Liposome

For nearly four decades, liposomes have been recognized as possible drug delivery vehicles (Nasr-Eldahan et al. 2021). The size is between 100 and 400 nm. Some liposome nanoparticle formulations include liposome-polycation-DNA nanoparticles and interlayered, crosslinked multilamellar vesicles. A liposome is a spherical vesicle made artificially from biologically inert lipids that are non-toxic and biodegradable and composed of a lamellar lipid bilayer. The antigen can be encapsulated within the core of liposomes, which are made up of biodegradable and harmless phospholipids (Giddam et al. 2012). The lipid bilayer structure facilitates the loading of both hydrophobic and hydrophilic compounds. Liposomes are phospholipid vesicles that develop spontaneously in aqueous solutions and have the ability to capture dissolved particles. They are biodegradable, slowly releasing the charged molecule as they decompose in the body. Depending on their size, the number of lamellae that make them up, and their ability to trap molecules in solution, they have diverse features. Given the high quantity of mucin in fish gills, the liposome charge must be taken into account while giving chemicals to fish in liposomes. The pH of the water deprotonates the mucin's sialic acid, enabling its interaction with cationic liposomes of <100 nm that contain DNA, for example, enhancing the residence time and uptake of the load. Because the interaction with the gills can produce hypoxia in the fish, this interaction results in large quantities of cationic liposomes, which are lethal to the fish. Similar amounts of anionic or neutral liposomes, on the other hand, are not lethal.

At present, liposomes are widely used as vaccine delivery vehicles in nanomedicine. Oral nanoliposomes have been used for fish vaccine administration with liposome nanoparticle-entrapping *A. salmonicida* and koi herpesvirus, which provided a better immune response than other formulations. A study was published on the effectiveness of nanoliposomes in the intraperitoneal injection of a *Vibrio harveyi* vaccine (Harikrishnan et al. 2012a, b, c). Phosphatidylcholine liposomes encapsulating inactivated *A. salmonicida* with formalin, as well as lipopolysaccharide (LPS) and inactivated toxin, were given to rainbow trout via immersion, which gave minimal protection against furunculosis, being slightly more efficient than the free antigen. Fernandez-Alonso et al. (1999) showed that when liposomes are given to fish, they can enable DNA expression. The green fluorescent protein (GFP) was detected in the fins of 0.2–0.5 g rainbow trout after immersion in 10–20 μm of DOTAP liposomes carrying codifying DNA for GFP (Fernandez-Alonso et al. 1999). There is no direct relationship between the size of liposomes and the organs in which they concentrate, according to studies, but there is a relationship in terms of their ability to lodge in specific organs. When rainbow trout were given large unilamellar phosphatidylcholine liposomes (LUV, 250 nm), they accumulated in larger proportions in their organs than multilamellar liposomes (MLV, 1–5 μm). They were collected (in decreasing order) in the spleen, head kidney, posterior kidney, visceral fat, and liver 24 h after treatment; liposome accumulation in hematopoietic organs has also been seen. Because liposomes are easily destroyed in the stomach, trials of fish vaccines using liposomes have been used for intraperitoneal injection or immersion delivery. More research into different forms of liposomes is needed to establish their use in the encapsulation of antigens for fish. Oral vaccinations are now possible thanks to modified liposomes that are resistant to stomach digestion.

Certain factors like net charge, lipid composition, particle size, and amount of loaded compound determine the vaccine delivery potential of the liposomes. Especially, the net charge of liposomes indirectly contributes to acute toxicity in fish after treatment due to the presence of high levels of mucin in fish gills, resulting in an unintended interaction between cationic liposomes and anionic mucin. However, a study investigated the effect of oral immunization with liposome-entrapped bacterial antigen on protection against *A. hydrophila* (a bacterial pathogen majorly attacking mucosal surfaces in the fish intestine and causing furunculosis disease). For the nanovaccine formulation, the liposome carrier was synthesized by combining dipalmitoylphosphatidylcholine, dipalmitoylphosphatidylserine, and cholesterol, followed by entrapping the *A. hydrophila* antigen in saline solution. The liposome-entrapped *A. hydrophila* was orally administered in *Cyprinus carpio* (25–30 g). A comparison of immunized with non-immunized fishes showed an 83.5% survival rate during pathogen invasion (Choi and Oh 2007). Similarly, another study conducted liposome (made of dipalmitoylphosphatidylcholine, dipalmitoylphosphatidylserine, cholesterol)-based oral vaccine administration in *E. bruneus* (29.5 ± 2.1 g) against *Vibrio harveyi* in kelp grouper and showed 90% mortality in non-immunized fish during *V. harveyi* infection (Harikrishnan et al. 2012a, b, c). Thus, the potential of the liposomal conjugated bacterial vaccine showed tremendous potential in fish survival. Further, a newer strategy to protect fish against viral/bacterial infection by immunostimulation of *Danio rerio* (0.6 ± 0.12 g) with nanoliposome co-encapsulating poly (Inosinic: Cytidylic) (a synthetic analogue of viral double-stranded RNA) and bacterial LPS to protect against lethal virus spring viraemia of carp virus and bacteria *P. aeruginosa* (PAO1) showed promising avenues in fish immunization (Ruyra et al. 2014).

## 3.6 Nanoemulsion

Nanoemulsions usually have a diameter of 20–200 nm and are an isotropic system made up of two immiscible liquids (water and oil) that stabilize with the addition of a surfactant. They can be in the form of water in oil or oil in water, with vaccinations in their cores, or simply mixed with antigens for delivery. Some emulsion-based nanoparticles including MF59 and Montanide are tailorable nano-sized emulsions. Polymeric chitosan and PLGA nanoparticles have been the most studied nanoparticles in fish vaccination research so far.

### *3.6.1 Immunostimulating Complex (ISCOM)*

Immunostimulating complexes contain immune-stimulating qualities, and they are commonly employed as a vaccination adjuvant to boost the immunological response and provide extended protection. The ISCOMs are spherical structures with an open cagelike structure (40 nm in diameter) that developed spontaneously when cholesterol, phospholipids, and quillaia saponins were mixed in a certain stoichiometry (Aguila et al. 2006). They have excellent adjuvant activity against a wide spectrum of bacterial and viral antigens. Virus-like particles are biocompatible capsid proteins that self-assemble into nanoparticles. They are great nanovaccines because they lack infectious nucleic acids but retain the virus's developed shape, which induces immunity. The size of nanoparticles utilized in vaccine development is usually between 20 and 800 nm. Immunostimulant complexes have been studied for more than three decades and are only available for veterinary use due to their low toxicity and haemolytic qualities (Smith et al. 2015). Although different forms of saponin have been examined as adjuvants in fish vaccines, there is just one research on the nano form (ISCOMs) for vaccine administration. In the study, the main OMPs of *A. hydrophila* were encapsulated in ISCOMs and administered intraperitoneally to eels, which provided good protection.

## 3.7 Immune Responses Due to Nanovaccines

The immune system of fish is divided into two types: innate and adaptive. Surface barriers (mucus, skin, gills, gastrointestinal tract), growth inhibitors (transferrin, interferon), enzyme inhibitors, and other innate defence mechanisms in fish are activated rapidly after infection. Nonspecific cellular factors are phagocytes, lysins (complement, antimicrobial peptides, lysozyme), precipitins and agglutinins (pentraxins, lectins), macrophages and neutrophils, phagocyte-activating chemicals (opsonins, cytokines), natural cytotoxic cells, eosinophils, basophils, mast cells, and inflammation. Adaptive immune responses take several days to become active, after which they provide basic memory cells, which are required for full pathogen elimination. Humoral immunity, cell-mediated immunity, and immunological memory are three facets of the adaptive immune system mediated by lymphocytes. Humoral immunity is characterized by the production of immunoglobulins (Ig) by B-cells, and there are three types of Igs identified in fish to date (IgM, IgD, and IgT) (Ballesteros et al. 2013). Cytotoxic T-lymphocytes are an integral part of the cellular immune system. Pathogens are recognized by adaptive immunity by molecules generated by somatic pathways, which are then followed by humoral and cellular reactions mediated by B- and T-lymphocytes. Dendritic cells, for example, are APCs that play a key role in both innate and adaptive immune responses. The APCs mature in response to microbial surface determinants, resulting in the transfer of MHC molecules (MHC I and MHC II) from intracellular compartments to the cell surface, the secretion of cytokines, morphological changes in dendritic cells, and cytoskeleton reorganization. The antigens are either internalized through the endocytic or non-endocytic pathways. The endocytic pathways, for example, entail the phagocytosis of antigen by APCs and the subsequent degradation of the antigen by proteolytic enzymes and reactive oxygen species. The degradation products (peptides) are then displayed on MHC class II molecules and identified by CD4+ T cells, causing antibody production and memory T-cell development. Non-endocytic pathways: Pathogen antigens are digested by the proteasome, which then displays peptides on MHC class I molecules (Shen et al. 2006). The CD8+ T cells that have cytotoxic activity against infected host cells identify the displayed antigen. The expression of co-stimulatory molecules (maturation markers) increased in a dose-dependent manner after these dendritic cells (DC) were exposed to nanoparticles (Uto et al. 2009). For the induction of DC maturation, however, both the absorption of nanoparticles and the characteristics of the polymers that shape the nanoparticles are critical. Despite their smaller size, nanoparticles have a greater influence on DC activation. As a result, surface interactions between nanoparticles and DCs influence DC maturation. When DCs were matured by PLGA nanoparticles, the expression of MHC class II and CD86 increased slightly compared to controls (Elamanchili et al. 2004). Nanoparticles elicit a variety of immune responses when administered, but they are not immunogenic unless they have been conjugated with an antigen. Pattern recognition receptor activation, cytotoxic T-lymphocyte induction, T-helper (Th) activation, cytokine development in different forms, B-cell activation, and antibody production are all involved in the induction of immune responses through various nanoparticles (Najafi-Hajivar et al. 2016). The size of the particles may play a role in the type of immunity that is caused. The APCs pick up nanoparticles depending on their size (Fifis et al. 2004). Smaller particles induce stronger immune responses than larger particles, according to several studies (Manolova et al. 2008).

### *3.7.1 Nanoparticle-Antigen Interaction*

Nanoparticles can deliver antigens to the immune cell in two ways: (a) immune cell co-ingestion of antigen and nanoparticle and (b) transient transmission, i.e. protecting the antigen and controlling its release at the target site. Nanoparticles engage certain immunological pathways in immune potentiator techniques, accelerating antigen processing and increasing immunogenicity (Mody et al. 2013). Simple physical adsorption or more nuanced methods such as encapsulation or chemical conjugation have been used to bind antigens. The charge or hydrophobic interaction is used to physically adsorb antigen onto a nanoparticle (Wendorf et al. 2006), where the contact between the nanoparticle and antigen is rather weak, resulting in quick disassociation *in vivo*. Encapsulation and chemical conjugation of antigens to nanoparticles provide the strongest interactions. During formulation, antigens are combined with nanoparticle precursors, resulting in antigen encapsulation in nanoparticles (Zhao et al. 2014). The antigen, on the other hand, is chemically cross-linked to the surface of a nanoparticle, and after being taken up by the nanoparticle, it is released inside the cell (Slütter et al. 2010). Immune potentiator techniques do not require antigen attachment or interaction with nanoparticles, and in some cases, they may be unfavourable in circumstances where the antigenic structure at the nanoparticle contact is altered. Several studies have shown that unique antibodies can be generated against nanoparticles, which is not a desirable trait because it could reduce the efficacy of nanovaccines (Zolnik et al. 2010; Zaman et al. 2013). Nanoparticles are not antigenic by themselves, but they do have antigenic properties when conjugated with antigens (proteins) due to their larger size (Zolnik et al. 2010; Zaman et al. 2013). T cells' activity in combating infections and cellular components is called cell-mediated immunity, and it's critical for protecting against a variety of pathogens. A significant cellular immune response can be induced by giving nanoparticle-based vaccinations in non-fish models, according to several studies (Zaman et al. 2013). Immunological memory is an essential feature of a specific immune response, which includes adaptive changes in lymphoid cells. The immune system identifies and kills a pathogen when it is exposed, which is the basis for an effective vaccine strategy (Fig. 3.2).

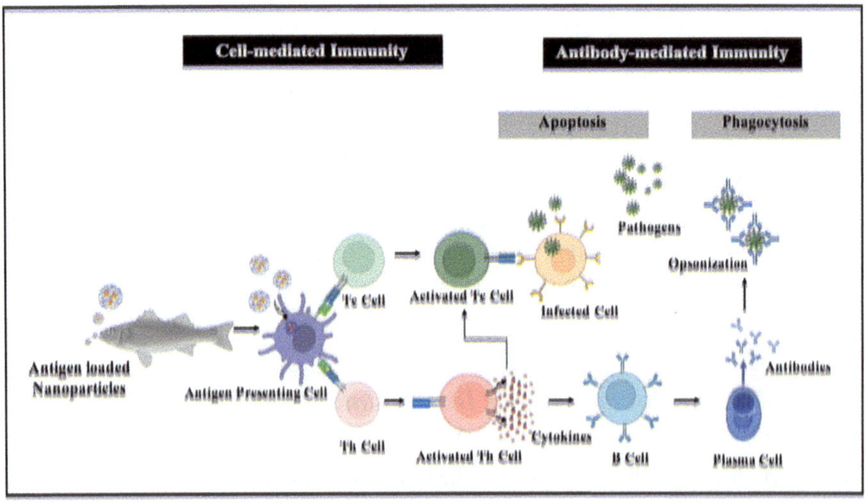


**Fig. 3.2** Adaptive immune response against nanovaccines within a fish. Created with BioRender.com

The production of a vaccine necessitates the activation of both the innate and adaptive immune systems, which function in tandem. Furthermore, adjuvants and delivery systems are needed to enhance and prolong the immune response, and various nanoparticles have emerged as frontrunners due to their unique properties. Nanoparticle-based vaccines help to close the gap by inducing the upregulation of many inflammatory, innate, and basic immune-sensitive genes (Zhu et al. 2014; Zheng et al. 2016). In fish, nanovaccines can elicit cellular and humoral immune responses, but cellular responses remain elusive. Based on the identification of the antigen by different methods, it has been suggested that oral nanovaccines can efficiently deliver purified antigens or DNA vaccines in the intestine, gills, liver, muscle, heart, blood, spleen, and kidney (Vimal et al. 2012). Several studies have shown that NPs can target the liver and other organs. Some nanovaccines can alter cell junction integrity during the permeation process, a mechanism that has been linked to the positive charges of NPs and is proposed as a useful property for enhancing antigen or DNA vaccine delivery (Liu et al. 2016). Fish innate immune cells respond to oral nanovaccines by increasing respiratory burst activity and immune-related enzymatic activities such as lysozyme, myeloperoxidase, and superoxide dismutase (Kole et al. 2018). Since systemic and mucosal (skin)-specific antibodies have been identified, B cells are directly involved (Rivas-Aravena et al. 2015). Except for TLR22 and NOD1 receptors, the function of receptors in vaccinated fish has been studied in depth (Liu et al. 2016; Kole et al. 2018). Immune-related gene expression research has revealed new information about the immunological pathways linked to nanovaccines. The upregulation of genes linked to innate and adaptive immune responses, such as iNOS, IL-1β, TNF-α, Mxs, IFNs, IL-10, IL-12, TGF-β, MHC, CD4, and CD8, was discovered during the evaluation of oral nanovaccines in fish (Rivas-Aravena et al. 2015; Liu et al. 2016). These genes were chosen based on the predicted immunological responses following the administration of a particular nanovaccine.

### 3.7.2 Nanoparticle-Antigen-Presenting Cells

The encapsulation of antigenic cells in a nanoparticle has sparked research into the mechanism for efficient antigen transmission to APCs, which leads to antigen maturation and then cross-presentation to induce an immune response. The DCs prefer virus particles with a diameter of 20–200 nm, while macrophages prefer particles with a diameter of 0.5–5 nm. When the particle size was smaller, a higher proportion of DCs interacted with the polystyrene spheres. Similarly, macrophages ingested PLA nanoparticles with a diameter of 200–600 nm more effectively than microparticles (Kanchan and Panda 2007). Particle shape, size, and surface charge are all essential particulate physicochemical factors that influence how particles interact with APCs. In contrast to hydrophilic particles, hydrophobic particles have been shown to elicit a stronger immune response (Hillaireau and Couvreur 2009).

## 3.8 Benefits of Nanovaccines

In recent years, rapid advances in nano-sciences and nanotechnologies have opened up new frontiers for several industrial and aquaculture areas. Nanotechnology is undeniably a significant prospect for the economic and long-term development of aquatic resources in many countries. In aquaculture, nanotechnology has evolved into a comprehensive tool for addressing a wide range of issues. Nano-vaccination is a novel approach to improving vaccine immunogenicity by employing nanoparticles as a carrier or adjuvant. There are a variety of materials available today that can be employed in an antigen delivery system for oral fish immunization. The nanomaterial-based vaccine's advantages include targeted antigen delivery, antigen stability, improved release kinetics, higher immune protectiveness, and immunogenicity. Because they are biodegradable and biocompatible, as well as less toxic, they are regarded as a viable alternative to standard vaccines. Nanoparticles also serve as adjuvants, assisting in the stimulation of immune responses while also protecting the antigen from degradation and allowing for regulated antigen release at the targeted site. It provides gastrointestinal stability, which is a key requirement for oral vaccination. Nanoparticles can imitate a natural illness, reducing the requirement for a booster dose while also improving vaccine efficacy. Various biodegradable polymeric particles, such as PLGA or PLA, operate as adjuvants and aid in the establishment of long-lasting immunity following a single injection. Biodistribution is aided by surface modification. The NPS can also be combined with targeting ligands, allowing particles to be directed to specific cells or regions. Nanoparticle-based delivery of DNA vaccines to APCs is one of the most promising delivery technologies for optimizing DNA vaccine formulation for immunotherapy (Rauta and Nayak 2015). The advantages and disadvantages of various types of nanoparticles are summarized in Table 3.2.

Vaccination with nanoparticles has several advantages over traditional vaccines.

1. A nanoparticle's size and shape can be modified to imitate a pathogen, allowing for efficient lymphatic drainage and subsequent internalization in APCs.
2. In physiological settings, nanoparticles effectively prevent the encapsulated antigen from destruction.
3. The charge and size of the nanoparticle influence the particle's biodistribution and retention in lymph nodes and spleens, encouraging memory immune responses.
4. Antigen and adjuvant co-delivery through nanoparticles to specific APCs, resulting in optimum antigen presentation and immune activation.
5. NP systems designed to facilitate endosomal escape can transfer antigens to the cytosol of APCs, enabling efficient antigen cross-presentation and the production of cytotoxic CD8+ T lymphocyte responses.
6. Antigen and adjuvant co-delivery through nanoparticles to specific APCs, resulting in optimum antigen presentation and immune activation.
7. In aquaculture, oral administration is the easier and less expensive way to achieve antigen delivery since it eliminates the labour-intensive, costly, and inconvenient

**Table 3.2** Advantage and disadvantages of different types of nanoparticles

| Nanoparticle type | Advantage | Disadvantage |
|---|---|---|
| Polymeric nanoparticles | Biodegradability | Low aqueous stability |
| | | Low antigen loading capacity |
| | | Premature release of antigen |
| | Better immunogenicity | Insufficient antigen protection |
| | Targeted antigen delivery | |
| Inorganic nanoparticles | Easy to modify | Low biodegradability |
| | Better protection of absorbed antigens and less chance of premature release | Low aqueous solubility |
| Nanoemulsions | Possess self-adjuvant properties | Poor gastrointestinal stability |
| | Encapsulates both hydrophilic and hydrophobic antigens | Premature release of antigen |
| ISCOMS | Easy to encapsulate antigen | Lack of reproducibility |
| | Built-in adjuvant property of Quil A | |
| Nanoliposomes | Stable in gastrointestinal fluids when modified | Limited antigen loading capacity |
| | | Low mucous penetration |
| | Possess intrinsic adjuvant properties that accommodate both hydrophilic and hydrophobic antigens | Poor gastrointestinal stability of naked liposomes |
| Viruslike particles | High gastro-intestinal stability that mimics the original virus | Lack of reproducibility |
| | Possess self-adjuvant properties | Premature release of antigen |

injection technique. The synthesis of nanoparticles costs much less than any other adjuvant, and because they are thermostable, they do not require a cold chain for storage.

Recent research has found that encapsulating *A. hydrophila* OMPs in PLGA and PLA nanoparticles improved the severity and duration of the immune response. In common carp, zebrafish, and rohu, even single-walled carbon nanotubes improve the immune-protectiveness of DNA and act as a delivery route for recombinant proteins targeting specific diseases. Because nanotubes may infiltrate APCs and carry and translocate bioactive molecules, they are useful carriers of antigens (Liu et al. 2016).

## 3.9 Current and Future Challenges

In many countries, nanotechnology presents a significant possibility for the economy and the long-term development of aquatic resources. Although the use of nanotechnology in aquaculture is still in its early stages, it has the potential to solve the majority of the problems in the aquaculture and fisheries sectors with improved technical innovation at various levels. Nanoparticles have certain unique properties and have shown promising use in fish vaccine administration, but they also have the drawback of being able to cross the blood-brain barrier, which could cause major problems (Joyappa et al. 2009). The NP-based targeting and delivery system has a small size, and a large surface area can cause aggregation, making physical handling problematic. Other difficulties with employing these nanoparticles include a lack of knowledge about NP distribution and the unpredictability of the process. Nanoparticle toxicity raises biosafety problems. If nanoparticles are made of innocuous chemicals, they are not always hazardous. The principal targets of nanoparticles for immunotoxicity have been identified as cell-mediated immunity and phagocytic cells. Lysosomal instability, frustrated phagocytosis, and changes in phagocytic cell function are all signs of toxicity. Although the humoral immune system is less susceptible to direct nanoparticle immunotoxicity, it is essential for the nanoparticles' dispersion throughout the body and presentation to phagocytic cells. However, there is a lot of scientific confirmation and research that must be done in this area. Despite the wide range of nanomaterials available, most fish nanovaccines have been produced using polymeric nanoparticles (PLGA, chitosan, and nano-polyplexes). An investigation is required on a wide range of nanomaterials, which can lead to the field's expansion. As a delivery vehicle, a variety of materials, including metallic and other organic NPs, are being considered. Other NPs currently being developed for vaccines and tested in fish via parenteral or immersion methods include fundamental investigations in fish that are required to establish relationships between the physicochemical qualities of NPs and their stability, biodistribution, destiny, and ultimately efficacy. Although most vaccine prototypes have low toxicity, this area deserves more research, taking into account particle size, bio-inertness, biodegradability, and safe excretion. High concentrations of chitosan (20 g/ml) and PLGA (1.25 mg/ml) NPs in water, for example, cause lead toxicity in zebrafish (*D. rerio*) embryos, causing impairment of hatching or survival (Nikapitiya et al. 2018). These results may be linked to the agglomeration of NPs on the surface of the chorion and the subsequent induction of hypoxia. Biocompatibility, permeation potential, and interaction-mediated mechanisms of nanovaccines have all been studied *in vitro* using fish cell lines. Surprisingly, *in vivo* follow-up experiments have shown the biodistribution of oral nanovaccines, revealing that they end up in the stomach, blood, gills, kidney, spleen, and muscle. Studies on the antigen release rate, cellular uptake dynamics, and intracellular destinies should be added to this information (Dubey et al. 2016a, b). Despite their widespread use, there are questions about a few nanoparticles that exhibit varying degrees of toxicity. Inorganic carbon nanotubes are non-biodegradable and have been confirmed to be toxic (Mutlu et al. 2010). Unregulated applications and toxicity reports from a small number of *in vitro* studies can skew the public's perception of nanoparticles, causing unnecessary concern and casting doubt on the science of nanomedicine (Yildirimer et al. 2011). With the increasing number of nanoparticle applications in recent years, the mechanisms will become clearer, and perceptions can shift in either direction. Oral nanovaccine prototypes have been tested in fish against viral and bacterial diseases, with evidence demonstrating their defensive ability against six viruses and five bacterial organisms. As a result, the list of diseases for which this technology can be used must be expanded. There were no studies on fish nanovaccines for parasitic diseases. Vaccine candidates can be developed using plasmid DNA or purified antigens with proven efficacy. Nanoparticles'antifungal and antiviral properties against fish diseases must be investigated. Only a few studies have looked at the use of nanoparticles in the diagnosis of bacterial and fungal diseases in aquaculture. Given nanoparticles' demonstrated ability, more focused investigations of their use in many fish medicine research topics are needed to encourage more effective fish disease diagnostics and therapy (Danhier et al. 2012). Since the former mimics a natural infection, it would better reflect vaccine efficacy in a real scenario; challenge studies focused on experimental infection rather than intraperitoneal pathogenic challenges are required to test nanovaccines more accurately (Rombout et al. 2011) and extend the challenge duration after mucosal vaccination. This will allow researchers to determine whether the defence is a result of innate or adaptive immune responses. When developing mucosal vaccines for fish, it's important to remember that a variety of factors influence the induction of long-lasting adaptive immune responses, including age, genetics, and climate (Embregts and Forlenza 2016). Mechanistic experiments on oral fish nanovaccines show that they boost respiratory blast, myeloperoxidase, lysozyme, and superoxide dismutase activities, as well as humoral responses, as measured by total and specific antibody output in serum and skin mucus (Zheng et al. 2016). To better understand the behaviour of oral nanovaccines, a study of encapsulation efficiency and release of encapsulated antigens should be done on a case-by-case basis, and induced immune responses should be compared among fish species. The latest progress on oral nanovaccine prototypes for use in fish has shown that they are feasible for use in aquaculture. The following scientific breakthroughs were discovered: (i) Only organic NPs were used, with chitosan and PLGA serving as the most popular nanomaterials; (ii) plasmid DNA and purified recombinant antigen were chosen; (iii) an encapsulated strategy was chosen over a surface-displayed strategy; (vii) most oral nanovaccines improved survival relative to other vaccines and routes of administration; (v) nanopolyplexes have only been tested against a bacterial disease, while (vi) viruslike particles have only been investigated against viral pathogens; (vii) most oral nanovaccines improve survival as compared to other vaccines and administration methods; and (viii) nanovaccines must be manufactured and tested on a large scale. These breakthroughs pave the way for new research into the use of oral nanovaccines in fish aquaculture, which would benefit both the scientific community and the industry (Benezra et al. 2011). Several obstacles remain, including the difficulty of synthesizing non-aggregated nanoparticles with consistent and desirable properties, a lack of understanding of how the physical properties of nanoparticles affect their biodistribution and targeting, and how these properties influence their interactions with the biological system at all levels, from the cell to the tissue. As a result, rational design combined with the reproducible production of nanoparticles with desirable properties, functionalities, and efficacy will become increasingly necessary, and it is expected that the introduction of new technologies, such as microfluidics, for the regulated synthesis of nanoparticles, will speed the creation of suitable nanoparticles for pharmaceutical applications. Novel vaccine systems for unmet needs, such as single-dose and needle-free delivery, will become feasible soon by combining some other appealing properties, such as slow-release, targeting, and alternative administration methods and delivery pathways (Zhao et al. 2014). Concerns about the toxicity of the particles, as well as difficulties in producing the materials and presenting antigens in their native form, are all drawbacks of using NPs for vaccine delivery. As a result, rational design combined with a repeatable synthesis of nanoparticles with desirable properties, functionalities, and efficacy is becoming increasingly important. The introduction of emerging technologies is expected to accelerate the production of appropriate nanoparticles for pharmaceutical applications. This sector, however, is still in its early stages. It's important to dig deeper into the physical-chemical properties of the nanoparticles used in vaccine production, as well as the properties of the antigens after they've been encapsulated. Finally, to better understand the effects of oral nanovaccines, a study of the mediated immune responses should be conducted across fish species. There is still a significant research gap, and new types of efficient nanoparticles, such as dendrimer nanocapsules, mesoporous nanoparticles, and others that are now accessible, must be investigated to develop effective vaccine delivery systems for aquaculture organisms. Since the precise mechanism of action of nanoparticles has yet to be fully known, there is still concern about toxicity. Recent developments and additional research into the biocompatibility of these nanoparticles can alter perceptions, opening up new avenues for combating deadly pathogens in aquaculture.

## 3.10 Conclusions

Most biological macromolecules that are used as antigens in vaccines are not able to get through the biological membranes to elicit a potent immune response. Hence, nanovaccines can ensure effective vaccine delivery. Immunization through nanovaccines provides better targeting and stimulates an antibody response at a cellular level. Nanoparticle-encapsulated vaccines aid in a sustained release of the antigen, reducing the dosing frequency. By designing an effective oral feed formulated with nano-carriers, the stress induced by the injection system can be reduced. A controlled and sustained release of the antigen will ensure a good immunogenic memory. With these properties, nanovaccines have shown promising potential for the prevention of diseases in aquaculture. Hence, exploring new possibilities to increase nanovaccine safety and efficacy should be the chief objective for further research in the field to enable aquaculture industries all around the world to use nanovaccines routinely in the future.

**Acknowledgments** This study was supported by the Department of Science and Technology (DST), Government of India, through the Indo-Norway joint project (INT/NOR/RCN/BIO/P-01/2018) under Bioeconomy.

## References


Abdelghany SM, Quinn DJ, Ingram RJ et al (2012) Gentamicin-loaded nanoparticles show improved antimicrobial effects towards *Pseudomonas aeruginosa* infection. Int J Nanomedicine 7:4053

Adams A (2019) Progress, challenges and opportunities in fish vaccine development. Fish Shellfish Immunol 90:210–214

Adomako M, St-Hilaire S, Zheng Y, Eley J, Marcum RD, Sealey W, Donahower BC, LaPatra S, Sheridan PP (2012) Oral DNA vaccination of rainbow trout, *Oncorhynchus mykiss* (Walbaum), against infectious haematopoietic necrosis virus using PLGA [poly (D, L-lactic-co-glycolic acid)] nanoparticles. J Fish Dis 35:203–214

Aguila A, Donachie AM, Peyre M et al (2006) Induction of protective and mucosal immunity against diphtheria by a immune stimulating complex (ISCOMS) based vaccine. Vaccine 24:5201–5210

Aklakur M, Asharf Rather M, Kumar N (2016) Nanodelivery: an emerging avenue for nutraceuticals and drug delivery. Crit Rev Food Sci Nutr 56:2352–2361

Arca HC, Günbeyaz M, Senel S (2009) Chitosan-based systems for the delivery of vaccine antigens. Expert Rev Vaccines 8:937–953

Badekila AK, Kini S, Jaiswal AK (2021) Fabrication techniques of biomimetic scaffolds in three-dimensional cell culture: a review. J Cell Physiol 236:741–762

Ballesteros NA, Castro R, Abos B et al (2013) The pyloric caeca area is a major site for IgM+ and IgT+ B cell recruitment in response to oral vaccination in rainbow trout. PLoS One 8:e66118

Behera T, Swain P (2011) Antigen adsorbed calcium phosphate nanoparticles stimulate both innate and adaptive immune response in fish, *Labeo rohita* H. Cell Immunol 271:350–359

Benezra M, Penate-Medina O, Zanzonico PB et al (2011) Multimodal silica nanoparticles are effective cancer-targeted probes in a model of human melanoma. J Clin Invest 121:2768–2780

Bhardwaj P, Bhatia E, Sharma S et al (2020) Advancements in prophylactic and therapeutic nanovaccines. Acta Biomater 108:1–21

Bhavsar MD, Amiji MM (2007) Polymeric nano-and microparticle technologies for oral gene delivery. Expert Opin Drug Deliv 4:197–213

Chahal JS, Khan OF, Cooper CL et al (2016) Dendrimer-RNA nanoparticles generate protective immunity against lethal Ebola, H1N1 influenza, and toxoplasma gondii challenges with a single dose. Proc Natl Acad Sci U S A 113:E4133–E4142

Choi SH, Oh CH (2007) Effect of oral immunization with liposome-entrapped bacterial antigen on protection against experimental *Aeromonas hydrophila*. Int J Integr Biol 11:33–38

Collins C, Lorenzen N, Collet B (2019) DNA vaccination for finfish aquaculture. Fish Shellfish Immunol 85:106–125

Danhier F, Ansorena E, Silva JM et al (2012) PLGA-based nanoparticles: an overview of biomedical applications. J Control Release 161:505–522

Dong CF, Lin TL, Gong H, Ou YSD, Yang S. (2005) Major outer membrane protein (MOMP) of Aeromonas hydrophila induced protective immunity to European eels (*Anguilla anguilla*). Acta Hydrobiol Sin 29:285–290

Dubey S, Avadhani K, Mutalik S et al (2016a) *Aeromonas hydrophila* OmpW PLGA nanoparticle oral vaccine shows a dose-dependent protective immunity in rohu (*Labeo rohita*). Vaccine 4:21

Dubey S, Avadhani K, Mutalik S et al (2016b) *Edwardsiella tarda* OmpA encapsulated in chitosan nanoparticles shows superior protection over inactivated whole cell vaccine in orally vaccinated fringed-lipped peninsula carp (*Labeo fimbriatus*). Vaccine 4:40

Elamanchili P, Diwan M, Cao M, Samuel J (2004) Characterization of poly (D, L-lactic-co-glycolic acid) based nanoparticulate system for enhanced delivery of antigens to dendritic cells. Vaccine 22:2406–2412

Embregts CW, Forlenza M (2016) Oral vaccination of fish: lessons from humans and veterinary species. Dev Comp Immunol 64:118–137

Fernandez-Alonso M, Alvarez F, Estepa A et al (1999) A model to study fish DNA immersion vaccination by using the green fluorescent protein. J Fish Dis 22:237–241

Fifis T, Mottram P, Bogdanoska V et al (2004) Short peptide sequences containing MHC class I and/or class II epitopes linked to nano-beads induce strong immunity and inhibition of growth of antigen-specific tumour challenge in mice. Vaccine 23:258–266

Foged C, Brodin B, Frokjaer S, Sundblad A (2005) Particle size and surface charge affect particle uptake by human dendritic cells in an *in vitro* model. Int J Pharm 298:315–322

Fredriksen BN, Grip J (2011) PLGA/PLA micro-and nanoparticle formulations serve as antigen depots and induce elevated humoral responses after immunization of Atlantic salmon (*Salmo salar* L.). Vaccine 30:656–667

Fredriksen BN, Sævareid K, McAuley L et al (2011) Early immune responses in Atlantic salmon (*Salmo salar* L.) after immunization with PLGA nanoparticles loaded with a model antigen and β-glucan. Vaccine 29:8338–8349

Ganda IS, Zhong Q, Hali M et al (2017) Dendrimer-conjugated peptide vaccine enhances clearance of *chlamydia trachomatis* genital infection. Int J Pharm 527:79–91

Gheibi Hayat SM, Darroudi M (2019) Nanovaccine: A novel approach in immunization. J Cell Physiol 234:12530–12536

Giddam AK, Zaman M, Skwarczynski M, Toth I (2012) Liposome-based delivery system for vaccine candidates: constructing an effective formulation. Nanomed J 7:1877–1893

Halimi M, Alishahi M, Abbaspour MR et al (2019) Valuable method for production of oral vaccine by using alginate and chitosan against *Lactococcus garvieae/streptococcus iniae* in rainbow trout (*Oncorhynchus mykiss*). Fish Shellfish Immunol 90:431–439

Harikrishnan R, Balasundaram C, Heo MS (2012a) Effect of Inonotus obliquus enriched diet on hematology, immune response, and disease protection in kelp grouper, *Epinephelus bruneus* against *Vibrio harveyi*. Aquac Res 344:48–53

Harikrishnan R, Balasundaram C, Heo MS (2012b) Poly d, l-lactide-co-glycolic acid (PLGA)-encapsulated vaccine on immune system in *Epinephelus bruneus* against *Uronema marinum*. Exp Parasitol 131:325–332

Harikrishnan R, Kim JS, Balasundaram C, Heo MS (2012c) Vaccination effect of liposomes entrapped whole cell bacterial vaccine on immune response and disease protection in *Epinephelus bruneus* against *Vibrio harveyi*. Aquac Res 342:69–74

Heegaard PM, Boas U, Sorensen NS (2010) Dendrimers for vaccine and immunostimulatory uses. A review. Bioconjug Chem 21:405–418

Heidarieh M, Moodi S, Katuli KK, Unger H (2015) Biochemical effects of encapsulated radio-vaccine via alginate nanoparticles as useful strategy for booster in immunized rainbow trout against *Ichthyophytirius multifiliis*. Acta Sci Vet 43:1330

Hillaireau H, Couvreur P (2009) Nanocarriers' entry into the cell: relevance to drug delivery. Cell Mol Life Sci 66:2873–2896

Irie T, Watarai S, Iwasaki T, Kodama H (2005) Protection against experimental *Aeromonas salmonicida* infection in carp by oral immunization with bacterial antigen entrapped liposomes. Fish Shellfish Immunol 18:235–242

Jeong KH, Kim HJ, Kim HJ (2020) Current status and future directions of fish vaccines employing virus-like particles. Fish Shellfish Immunol 100:49–57

Ji J, Torrealba D, Ruyra À, Roher N (2015) Nanodelivery systems as new tools for immunostimulant or vaccine administration: targeting the fish immune system. Biology 4:664–696

Jovanovic B, Anastasova L, Rowe EW et al (2011) Effects of nanosized titanium dioxide on innate immune system of fathead minnow (*Pimephales promelas* Rafinesque, 1820). Ecotoxicol Environ Saf 74:675–683

Joyappa DH, Kumar CA, Banumathi N et al (2009) Calcium phosphate nanoparticle prepared with foot and mouth disease virus P1-3CD gene construct protects mice and Guinea pigs against the challenge virus. Vet Microbiol 139:58–66

Kalkanidis M, Pietersz GA, Xiang SD et al (2006) Methods for nano-particle based vaccine formulation and evaluation of their immunogenicity. Methods 40:20–29

Kanchan V, Panda AK (2007) Interactions of antigen-loaded polylactide particles with macrophages and their correlation with the immune response. Biomaterials 28:5344–5357

Katare YK, Panda AK (2006) Influences of excipients on *in vitro* release and *in vivo* performance of tetanus toxoid loaded polymer particles. Eur J Pharm Sci 28:179–188

Kavaliauskis A, Arnemo M, Speth M et al (2016) Protective effect of a recombinant VHSV-G vaccine using poly (I: C) loaded nanoparticles as an adjuvant in zebrafish (*Danio rerio*) infection model. Dev Comp Immunol 61:248–257

Khurana A, Tekula S, Saifi MA et al (2019) Therapeutic applications of selenium nanoparticles. Biomed Pharmacother 111:802–812

Klajnert B, Bryszewska M (2001) Dendrimers: properties and applications. Acta Biochim Pol 48:199–208

Kobayashi K, Wei J, Iida R et al (2014) Surface engineering of nanoparticles for therapeutic applications. Polym J 46:460–468

Kole S, Kumari R, Anand D et al (2018) Nanoconjugation of bicistronic DNA vaccine against *Edwardsiella tarda* using chitosan nanoparticles: evaluation of its protective efficacy and immune modulatory effects in *Labeo rohita* vaccinated by different delivery routes. Vaccine 36:2155–2165

Kole S, Qadiri SSN, Shin SM et al (2019) PLGA encapsulated inactivated-viral vaccine: formulation and evaluation of its protective efficacy against viral haemorrhagic septicaemia virus (VHSV) infection in olive flounder (*Paralichthys olivaceus*) vaccinated by mucosal delivery routes. Vaccine 37:973–983

Li L, Lin SL, Deng L, Liu ZG (2013) Potential use of chitosan nanoparticles for oral delivery of DNA vaccine in black seabream *Acanthopagrus schlegelii* Bleeker to protect from *Vibrio parahaemolyticus*. J Fish Dis 36:987–995

Liu Y, Wang FQ, Shah Z et al (2016) Nano-polyplex based on oleoyl-carboxymethy-chitosan (OCMCS) and hyaluronic acid for oral gene vaccine delivery. Colloids Surf 145:492–501

Maiti B, Shetty M, Shekar M et al (2012) Evaluation of two outer membrane proteins, Aha1 and OmpW of *Aeromonas hydrophila* as vaccine candidate for common carp. Vet Immunol Immunopathol 149:298–301

Maiti B, Dubey S, Munang'andu HM et al (2020) Application of outer membrane protein-based vaccines against major bacterial fish pathogens in India. Front Immunol 11:1362

Manolova V, Flace A, Bauer M et al (2008) Nanoparticles target distinct dendritic cell populations according to their size. Eur J Immunol 38:1404–1413

McCall RL, Sirianni RW (2013) PLGA nanoparticles formed by single-or double-emulsion with vitamin E-TPGS. J Vis Exp 82

Mody KT, Popat A, Mahony D, Cavallaro AS, Yu C, Mitter N (2013) Mesoporous silica nanoparticles as antigen carriers and adjuvants for vaccine delivery. Nano 5:5167–5179

Munang'andu HM, Fredriksen BN, Mutoloki S et al (2012) Comparison of vaccine efficacy for different antigen delivery systems for infectious pancreatic necrosis virus vaccines in Atlantic salmon (*Salmo salar* L.) in a cohabitation challenge model. Vaccine 30:4007–4016

Mutlu GM, Budinger GS, Green AA et al (2010) Biocompatible nanoscale dispersion of single-walled carbon nanotubes minimizes *in vivo* pulmonary toxicity. Nano Lett 10:1664–1670

Najafi-Hajivar S, Zakeri-Milani P, Mohammadi H et al (2016) Overview on experimental models of interactions between nanoparticles and the immune system. Biomed Pharmacother 83:1365–1378

Nasr-Eldahan S, Nabil-Adam A, Shreadah MA et al (2021) A review article on nanotechnology in aquaculture sustainability as a novel tool in fish disease control. Aquac Int 29:1–22

Nikapitiya C, Dananjaya SHS, De Silva BCJ et al (2018) Chitosan nanoparticles: a positive immune response modulator as display in zebrafish larvae against *Aeromonas hydrophila* infection. Fish Shellfish Immunol 76:240–246

Pati R, Shevtsov M, Sonawane A (2018) Nanoparticle vaccines against infectious diseases. Front Immunol 9:2224

Poobalane S, Thompson KD, Ardó L et al (2010) Production and efficacy of an *Aeromonas hydrophila* recombinant S-layer protein vaccine for fish. Vaccine 28:3540–3547

Prokop A, Davidson JM (2008) Nanovehicular intracellular delivery systems. J Pharm Sci 97:3518–3590

Rajesh Kumar S, Ishaq Ahmed VP, Parameswaran V et al (2008) Potential use of chitosan nanoparticles for oral delivery of DNA vaccine in Asian sea bass (*Lates calcarifer*) to protect from *vibrio (Listonella) anguillarum*. Fish Shellfish Immunol 25:47–56

Rauta PR, Nayak B (2015) Parenteral immunization of PLA/PLGA nanoparticle encapsulating outer membrane protein (Omp) from *Aeromonas hydrophila*: evaluation of immunostimulatory action in *Labeo rohita* (rohu). Fish Shellfish Immunol 44:287–294

Rivas-Aravena A, Sandino AM, Spencer E (2013) Nanoparticles and microparticles of polymers and polysaccharides to administer fish vaccines. Biol Res 46:407–419

Rivas-Aravena A, Fuentes Y, Cartagena J et al (2015) Development of a nanoparticle-based oral vaccine for Atlantic salmon against ISAV using an alphavirus replicon as adjuvant. Fish Shellfish Immunol 45:157–166

Rombout JH, Abelli L, Picchietti S et al (2011) Teleost intestinal immunology. Fish Shellfish Immunol 31:616–626

Ruyra A, Cano-Sarabia M, García-Valtanen P et al (2014) Targeting and stimulation of the zebrafish (*Danio rerio*) innate immune system with LPS/dsRNA-loaded nanoliposomes. Vaccine 32:3955–3962

Salonius K, Simard N, Harland R, Ulmer JB (2007) The road to licensure of a DNA vaccine. Expert Opin Investig Drugs 8:635

Salvador A, Igartua M, Hernández RM, Pedraz JL (2011) An overview on the field of micro-and nanotechnologies for synthetic peptide-based vaccines. J Drug Deliv

Semete B, Booysen L, Lemmer Y et al (2010) In vivo evaluation of the biodistribution and safety of PLGA nanoparticles as drug delivery systems. Nanomedicine 6:662–671

Shen H, Ackerman AL, Cody V et al (2006) Enhanced and prolonged cross-presentation following endosomal escape of exogenous antigens encapsulated in biodegradable nanoparticles. Immunology 117:78–88

Slütter B, Bal SM, Que I et al (2010) Antigen – adjuvant nanoconjugates for nasal vaccination: an improvement over the use of nanoparticles? Mol Pharm 7:2207–2215

Smith JD, Morton LD, Ulery BD (2015) Nanoparticles as synthetic vaccines. Curr Opin Biotechnol 34:217–224

Stride MC, Polkinghorne A, Nowak BF (2014) Chlamydial infections of fish: diverse pathogens and emerging causes of disease in aquaculture species. Vet Microbiol 170:19–27

Tian J, Yu J (2011) Poly (lactic-co-glycolic acid) nanoparticles as candidate DNA vaccine carrier for oral immunization of Japanese flounder (*Paralichthys olivaceus*) against lymphocystis disease virus. Fish Shellfish Immunol 30:109–117

Uto T, Akagi T, Hamasaki T et al (2009) Modulation of innate and adaptive immunity by biodegradable nanoparticles. Immunol Lett 125:46–52

Vartak A, Sucheck SJ (2016) Recent advances in subunit vaccine carriers. Vaccine 4:12

Vimal S, Taju G, Nambi KN et al (2012) Synthesis and characterization of CS/TPP nanoparticles for oral delivery of gene in fish. Aquaculture 358:14–22

Vimal S, Majeed SA, Nambi KSN et al (2014) Delivery of DNA vaccine using chitosan–tripolyphosphate (CS/TPP) nanoparticles in Asian sea bass, *Lates calcarifer* (Bloch, 1790) for protection against nodavirus infection. Aquaculture 420:240–246

Vinay TN, Tanmoy GC, Anutosh P et al (2016) Nanovaccines: a possible solution for mass vaccination in aquaculture. World Aquacult 31

Wang Y, Liu GL, Li DL et al (2015) The protective immunity against grass carp reovirus in grass carp induced by a DNA vaccination using single-walled carbon nanotubes as delivery vehicles. Fish Shellfish Immunol 47:732–742

Wendorf J, Singh M, Chesko J et al (2006) A practical approach to the use of nanoparticles for vaccine delivery. J Pharm Sci 95:2738–2750

Yasumoto S, Kuzuya Y, Yasuda M et al (2006) Oral immunization of common carp with a liposome vaccine fusing koi herpesvirus antigen. Fish Pathol 41:141–145

Yildirimer L, Thanh NT, Loizidou M, Seifalian AM (2011) Toxicology and clinical potential of nanoparticles. Nano Today 6:585–607

Zaman M, Good MF, Toth I (2013) Nanovaccines and their mode of action. Methods 60:226–231

Zhang L, Zeng Z, Hu C et al (2016) Controlled and targeted release of antigens by intelligent shell for improving applicability of oral vaccines. Biomaterials 77:307–319

Zhao L, Seth A, Wibowo N et al (2014) Nanoparticle vaccines. Vaccine 32:327–337

Zheng F, Liu H, Sun X et al (2016) Development of oral DNA vaccine based on chitosan nanoparticles for the immunization against reddish body iridovirus in turbots (*Scophthalmus maximus*). Aquaculture 452:263–271

Zhu B, Liu GL, Gong YX et al (2014) Single-walled carbon nanotubes as candidate recombinant subunit vaccine carrier for immunization of grass carp against grass carp reovirus. Fish Shellfish Immunol 41:279–293

Zhu B, Liu GL, Gong YX et al (2015) Protective immunity of grass carp immunized with DNA vaccine encoding the vp7 gene of grass carp reovirus using carbon nanotubes as a carrier molecule. Fish Shellfish Immunol 42:325–334

Zolnik BS, González-Fernández Á, Sadrieh N, Dobrovolskaia MA (2010) Minireview: nanoparticles and the immune system. Endocrinology 151:458–465

# Chapter 4
# Nanotechnology: A Novel Tool for Aquaculture Feed Development

**Digi George, Sreeja Lakshmi, Anuj Sharma, Sanchu Prakash, Muzammil Siddiqui, B. R. Malavika, and Preetham Elumalai**

## Abbreviations

| | |
|---|---|
| % WG | Percent weight gain |
| AgNPs | Silver nanoparticles |
| AgNPs | Silver nanoparticles |
| CD- $TiO_2$ | Carbon dots coupled with titanium dioxide |
| CD | Carbon dots |
| FCR | Feed conversion ratio |
| Fe | Iron |
| GIT | Gastrointestinal tract |
| GSH-Px | Glutathione peroxidase enzyme |
| Hb | Haemoglobin |
| IGF-1 | Insulin-like growth factor 1 |
| IgM | Immunoglobulin M |
| nFe | Iron nanoparticles |
| NPs | Nanoparticles |
| nSe | Selenium nanoparticles |
| $nTiO_2$ | Titanium dioxide nanoparticles |
| PL | Post larvae |
| RBCs | Red blood cells |
| ROS | Reactive oxygen species |

D. George · S. Lakshmi
School of Ocean Science and Technology, Kerala University of Fisheries and Ocean Studies, Panangad, Kerala, India

A. Sharma · S. Prakash · M. Siddiqui · B. R. Malavika
Kerala University of Fisheries and Ocean Studies, Kochi, Kerala, India

P. Elumalai (✉)
School of Ocean Science and Technology, Kerala University of Fisheries and Ocean Studies, Kochi, Kerala, India
e-mail: preetham@kufos.ac.in



V. K. Arivarasan et al. (eds.), *Nanotechnological Approaches to the Advancement of Innovations in Aquaculture*, Nanotechnology in the Life Sciences, https://doi.org/10.1007/978-3-031-15519-2_4

| | |
|---|---|
| Se | Selenium |
| SGR | Specific growth rate |
| $TiO_2$ | Titanium dioxide |
| WBCs | White blood cells |
| Zn | Zinc |
| ZnO | Zinc oxide |
| ZnO-NP | Zinc oxide nanoparticles |

## 4.1 Introduction

Aquaculture is an expanding area in the agriculture sector, contributing a total of 46%, i.e. 82 million tonnes in 2018. FAO 2020 reported around USD 250 billion from aquaculture production alone out of the total first-sale value of USD 401 billion from 179 million tonnes of global fish production. For billions of populations, this provides the best option for easily digestible protein and healthy fat, as well as many other essential micronutrients.

But many environmental issues are affecting the production of fish in water bodies. The extreme practice of aquaculture techniques is destructing the natural ecosystem through eutrophication, acidification, and chemical and biological contamination (Sarkar et al. 2021; Pudake et al. 2019). This results in an unsuitable environment for the rearing and breeding purposes of fish, which leads to the degradation of fish production. In this advanced world, the detection of contamination and toxicity levels in water bodies is tough and onerous work. Nutrients become unavailable in water bodies because of huge contamination by sewage inlets. Therefore, in an extensive aquaculture system, deficiency of nutrients has become dreadful, resulting in suboptimal breeding efficiency in juvenile and broodfish populations (Sarkar et al. 2021).

Nanotechnology can solve the recent problems in the agriculture and aquaculture sectors. It is the science of creating and using compounds with unique properties at nanoscale dimensions (1–100 nm) to enable new functionalities (Shaalan et al. 2016). The aquaculture industry can be transformed by nanotechnology through new tools for quick disease detection, nutrient delivery, improvement of the ability of fish to absorb vaccines and drugs, decreasing pollution, etc. (Pudake et al. 2019). In the aquaculture sector, nanotechnology has various practical purposes that help to improve management, such as wastewater treatment, fishpond sterilization, improving fish processing, avoiding nutritional deficiency, involvement in the feed industry, and health management. In this chapter, we focused on the role of nanotechnology in aquafeed development. The role of feed-in aquaculture has great importance because of its high input cost. The improvement in the fish feed will lead to high fish production, and nanotechnology has great possibilities to enhance the nutrient composition of feed, which will help to tackle the issues that originate from nutrient unavailability.

## 4.2 Role of Nanotechnology in Enhanced Feed Additive Preparation

Aquaculture has been fulfilling the demand for animal protein in a big way. The world population has been increasing, and so has its need for more food. Aquaculture has been contributing to the supply of animal protein, but as it gets more and more intensified, instances of stresses on the environment and on the fish species themselves are being witnessed. Diseases, harmful chemical accumulation, adverse effects on the ecosystem of the culture, incomplete and inadequate utilization of resources, faulty culture practices, incapable personnel involved in the culture, among others, limit the amount of produce we get through it. Nanotechnology is a branch of science that has very recently come into existence. It involves the study, use, manipulation, formation of particles that are in at least one of their dimensions between 1 and 100 nm. When bulky materials are changed to nanosize, then the surface chemistry and many other properties change. Nanomaterials are also naturally occurring in foods, such as proteins, fats, carbohydrates. They create their higher structures in our bodies. They have now been added intentionally also to our foods, feeds, etc. There are not many nanomaterials that are currently being applied as a feed additive, as it is a new field. General public perception on the fish produced on diets consisting of nanomaterials is still skeptical in many instances. Still, they are being researched and some are even substituting antibiotics (Peters et al. 2016). Nanomaterials may contain micronutrients. When fishes will consume them, the delivery of these nutrients will get precise. They can also change the density, shape, texture of the feed (Handy 2012). They will also improve the absorption and availability of nutrients and additives to the fish targeted (Chaudhry and Castle 2011). Due to their small size, the additives can be easily assimilated by cells and help in the rapid growth of fishes (Moges et al. 2020). Nanoparticles are prepared in several ways, which are influenced by the nature of the material, the stability of the material, its usage, etc. So, having nanotechnology as a tool to incorporate feed additives is possible. But risks that may be associated with the newly formulated materials must be considered. Safety evaluation is a must for them.

Nanoparticles can increase feed utilization by changing the colour, flavour, etc. Water-insoluble compounds such as some vitamins, minerals, etc. can be added to feeds, which makes them available to the fish (Sarkar et al. 2021). There is equipment available on the market that makes nanomaterials and adds them to the feed. Ubisol-AquaTM Delivery System Technology and NovaSOL developed by AQUANOVA are some of the examples which are creating the smallest of particles (Aklakur et al. 2016). Indian-origin companies are also producing equipment. Some of the examples are NannoCal Aqua (nano calcium), NanoPHOS (nano phosphorus), etc.

Nanoencapsulation is a process that protects the food from many factors, such as environmental stress, and masks inadequacies such as taste and odour. Also, it helps in bringing out the real taste of the feed (Fathi et al. 2012). Using different nanotechnology methods, lipids in solid form are also being encapsulated. Vitamin D and casein made micelles are made possible. Chitosan-based nanomolecules are enabling water-soluble entities such as ascorbic acid to also be capsulated (Jimenez-Fernández et al. 2014). Nanoparticles are in use in several shapes, which include nanospheres, nanotubes, nanocapsules, etc. (Shah and Mraz 2020). The shapes of the nanoparticles greatly affect their behavior. They can be elliptical, discoidal, conical, etc. (Bunglavan et al. 2014). Some of the common methods which are used to prepare nanoparticles are

- Cross-linking emulsion: A water-oil emulsion is made by vigorous shaking. Agents which can stabilize the product are also involved.
- Precipitation: Particles are produced by putting them in an alkaline solution. This process is also called 'blowing'. Purification is done by filtration and centrifugation, and then rinsing with cold and warm water.
- Spray drying: It is one of the easier processes. Drying is done. It is achieved by spraying them with compressed hot air. One chosen solvent is also used in this step, which evaporates instantly due to the high temperature of compressed air (Bunglavan et al. 2014).

A study was done in 2020 to determine the growth rate of silver catfish. Diphenyl diselenide nanocapsules were added to their feed. It resulted in the confirmation of the finding that growth was more in the group which had consumed the feed. In the experiment, $Ph_2Se_2$ added feed was produced using interfacial deposition of preformed polymers. For 1 hour, the organic component in it was kept at 40 °C temperatures in the water bath, then put into an aqueous phase and mixed for 10 minutes. Then, a rotatory evaporator was used to concentrate the mix and reduce the organic content. Later, the mixture was heated in a circulation oven for 24 hours, and the pellets were broken and stored (Baldissera et al. 2020).

Young carp showed faster growth when given iron nanoparticles in their feed. Nanoselenium is more potent than organic selenium for increasing selenium levels in muscles. It also improved the final weight, gain rate, antioxidant levels, etc. in the body (Sekhon 2014). Nutraceuticals are also being delivered using nanomaterials, which increase their potency, though the cost of feed preparation increases (Rather et al. 2011). Nanovaccines and nanoparticles used for gene therapy also involve the use of nanotechnology. In gene therapy, a carrier vehicle is designed to deliver DNAs (Sarkar et al. 2021). Compound Nano-863 is used in China for the growth and development of fish. They were produced by complementing nanomaterials exhibiting high-temperature sintering using a ceramic substance as a carrier and a strong light-absorbing capacity (Moges et al. 2020).

## 4.3 Types of Nanomaterial and Their Properties

According to their ability to relay various components and interact with diverse environmental circumstances, nanoparticles are categorized in various categories. Nanomaterials can be divided into four categories: carbon nanomaterials, inorganic-based nanomaterials, organic-based nanomaterials, and nanocomposites (Majhi and Yadav 2021; Khan and Khan 2020). Based on the chemical characteristics of the nanoparticle, it can be classified into inorganic and organic categories (FSAI 2008). The applications of nanoparticles are diverse, and they also help to improvise fish production in the aquaculture sector. It helps in disease detection, increases the drug absorption ability of fish, promotes vaccine development, upgrades water quality parameters, etc. (Pudake et al. 2019). The major cost in aquaculture is contributed by feed development, and nanoparticles help to improve the quality of feed by enhancing the nutrient composition, micronutrient delivery, feed encapsulation, and promoting the growth of fish. Some of the nanoparticles are categorized below based on feed improvements.

### 4.3.1 Inorganic Nanoparticles

These nanoparticles are manufactured by using inorganic ingredients at nanoscales already permitted for use in feed, e.g., the use of silver in poultry feed, which helps to improve the microbiota of chickens (Gangadoo et al. 2016). Metal-based inorganic nanomaterials comprise cadmium, silver, iron, gold, copper, aluminum, lead, and zinc nanomaterials, while titanium dioxide, iron oxide, cerium oxide, silica, zinc oxide, copper oxide, iron oxide, magnesium aluminum oxide, etc. are examples of metal oxide-based inorganic nanomaterials. Zinc helps to regulate energy consumption, metabolism of vitamin A and lipids, and protein synthesis when fed to animals along with feed (Pudake et al. 2019). nTiO2 is used to notice the growth and performance improvement of rainbow trout (Ramsden et al. 2009). It has been noticed that Nile tilapia (*Oreochromis niloticus*) and Prussian carp (*Carassius auratus gibelio*) fed on nSe as supplemented diets showed positive results such as a reduction in FCR with an increase in final weight and improved overall growth, respectively (Zhou et al. 2009; Deng and Cheng 2003).

### 4.3.2 Organic Nanoparticles

Organic-based nanomaterials are made up of organic compounds that do not contain carbon, such as liposomes, cyclodextrin, dendrimers, and micelles (Majhi and Yadav 2021). Organic nanoparticles are most likely utilized to enhance or modify food functionality to improve the nutritional value of food systems. These nanoparticles were aimed at providing vitamins and other nutrients without affecting the flavour or look of food and beverages. Such nanoparticles wrap the nutrients and transport them into the bloodstream via the gastrointestinal tract (GIT), enhancing their bioavailability (FSAI 2008). Micelles are organic nano particulates that an wrap nonpolar molecules like flavours, vitamins, antioxidants, lipids, and antimicrobials (Chen et al. 2006). Liposomes are used to facilitate the delivery of functional components to food such as antimicrobials, nutraceuticals, and flavouring properties (Peters et al. 2011).

### 4.3.3 Nanocomposites

Composite nanomaterials or nanocomposites are the amalgamations of carbon-based, metal-based, and metal oxide-based nanomaterials, and they have complex structures such as metal-organic structures (Majhi and Yadav 2021). These materials have been used by researchers to find various benefits of nanocomposites in the aquaculture sector. Abad-Álvaro et al. 2019 used two various clay nanocomposites (i.e. kaolinite-Ag and sepiolite-Ag) in which sepiolite and kaolinite act as carriers for silver nanoparticles (AgNPs) that are fed orally and found that only limited release of silver nanoparticles is done by carriers due to the formation of silver chloride during stomach stimulation (Abad-Álvaro et al. 2019). Various nanocomposites are also used during fish processing, and chitosan composites are one of them, which are used during the refrigeration of fish meat at 4 °C and help to inhibit bacterial growth and decrease the formation of volatile bases and oxidation products that spoil the fish during freezing (Ahmed et al. 2019). In aquaculture wastewater effluents, Louros et al. 2021 try to increase the photodegradation of antibiotics by solar irradiation using carbon dots (CD) coupled with titanium dioxide ($TiO_2$) (CD-$TiO_2$) and conclude that ecofriendly CD-$TiO_2$ hybrid materials photocatalyze efficiently and show sustainable as well as promising strategies to expedite the removal of antibiotics effectively from aquaculture effluents (Louros et al. 2021).

### 4.3.4 Carbon Nanomaterials

Carbon black, fullerene, multi-walled carbon nanotubes, graphene, single-walled carbon nanotubes, activated carbon, and carbon fiber are forms of carbon-based nanomaterials (Majhi and Yadav 2021). These materials can also be used to improve the quality of water. According to Baby et al. 2019, carbon nanomaterials can be used to absorb various harmful gases from the aquatic environment. It has been reported that Graphene, a type of carbon nanomaterial, can be used to capture $CO_2$ and $H_2$. The best-superior materials used for the elimination of organic contaminants from the water are based on graphene nano-adsorbents (Baby et al. 2019; Bradder et al. 2011).

Different types of nanomaterials have specific properties that help for good aquaculture management and maintain better animal health. Some of them are discussed in Table 4.1 along with their properties.

**Table 4.1** Various properties of nanoparticles

| Nanoparticles | Properties (as feed additives) | Reference |
|---|---|---|
| Silver | Antimicrobial agent in additives for animal feed<br>Food supplements<br>Act as nanocarriers to increase the absorption of nutrients in fish and shellfish | Abad-Álvaro et al. (2019), FSAI (2008), De Silva et al. (2021) |
| Nano-ZnO (nZnO) | Improve the specific growth rate (SGR), % weight gain, and feed conversion ratio (FCR)<br>Antimicrobial activity<br>Induced growth hormone level in serum<br>Improve the concentration of total nitrogen and immunoglobin M | Chris et al. (2018), Tawfik et al. (2017) |
| Iron | Food supplement<br>Modifies the appropriate performance of the central nervous system<br>Improve the SGR, overall weight gain, and FCR<br>Increase the level of IGF-1 in animals | FSAI (2008), Akbary and Jahanbakhshi (2019) |
| Copper | Improved body weight while improving FCR<br>Decrease in mortality rate<br>Improved growth and immunity performance | Gangadoo et al. (2016) |
| Selenium | Stimulate growth hormone production<br>Regulate thyroid hormone production in fish<br>Improve larval growth related to bone mineralization | Dawood et al. (2021) |

## 4.4 Application of Nanoparticles in the Aquaculture Sector

The aquaculture sector is one of the world's most developing domains with high growth promise (Sarkar et al. 2021). In the racing sector, where progress in science and technology has proved sustainable and efficient development, the use of nanoparticles is one of the many advancements in this sector. Particles with diameters ranging from 1 to 100 nm are considered nanoparticles. The particles in the nanosize range demonstrate new chemical and physical phenomena (Márquez et al. 2018). Nanoparticles can be found in individual or aggregate form. These aggregates can have a size that extends over 100 nm and thus don't fall under the category of nanomaterials, but these will still have the properties of nanoparticles (Shah and Mraz 2020). Various experiments and research proved that nanomaterials have a broad range of uses in the aquaculture sector (Márquez et al. 2018). They can be used for the control of disease in fishes, water treatment, sterilization of pond water, processed nanofood, vaccine transport, etc. Mostly metal nanoparticles such as silver, selenium, copper, gold, etc. are used.

The nanoparticle can provide notable benefits in the transport of nutrients needed in minor concentrations and unstable components to the vital organs of the fish. Fatty acids, other highly soluble nutrients, and components that have limited uptake efficacy can be encapsulated using nanomaterials to avoid disassociation in the gut of the fish (Handy 2012). The nanoparticle can provide trace metal to fishes without causing much loss of faecal matter, which is generally related to mineral salt. Various nanoparticles can perform as immunomodulators and development enhancers when supplied with aquafeed in smaller quantities. Selenium nanoparticles, when induced into Crucian carp along with the feed, showed an increase in weight, muscle development, and antioxidant characteristics. Research showed that nanoparticles of silver given to *Danio rerio* (zebrafish) enhanced growth and increased metalloprotease and protease activity (Sarkar et al. 2021). Nanoparticles when used to integrate vitamin C along with the feed for *Onchorhychus mykiss* (Rainbow trout) showed an enhancement in active permanency for about 20 days, whereas active permanency was lost in the initial 3 days when the fishes were given conventional feed. The nanoparticle was also able to secure vitamin C from various acids and enzymes in the gut of the rainbow trout as it showed prolonged production of vitamin C in the gut epithelium, which improves the nonspecific immune response of the fish (Márquez et al. 2018). When iron nanoparticles and *Lactobacillus casei* were provided along with feed for rainbow trout, they showed remarkable growth parameters (Shah and Mraz 2020). Selenium, manganese, and zinc nanoparticles given to *Sparus aurata* (gilthead sea bream) in the initial feed showed an increase in bone mineralization and tolerance to stress (Shah and Mraz 2020). Along with increasing the firmness and biological accessibility of the feed components, nanomaterials can also be used to change the physical parameters of fish feed. The instability, poor buoyancy, and texture of pelleted feed usually are primary causes of food wastage, which may lead to water quality deterioration, which is a major issue. Adding nanomaterial in small quantities can alter the physical parameters of feed pellets (Handy 2012).

Another important aspect of the nanoparticle is its ability to control disease (Márquez et al. 2018). A few biomaterials cannot attach to the disease-causing microbes because of their lower specificity and thus can't be used as a major target of the cells of disease-causing microbes; this leads to the use of nanoparticles (Munawar 2021). The silver nanoparticle has a bacteria-killing ability (Márquez et al. 2018). Nanoparticles of copper (mostly n-copper oxides) have the ability for antifouling and bactericide, and they are also a good conductor of heat (Khosravi-Katuli et al. 2017). Nanoparticles of n-iron oxides have properties such as lower toxicity and special surface properties, which can be used in the aquaculture sector. NPs of iron oxides can be used for the repair of tissues, delivery of medicine, and cellular labeling (Khosravi-Katuli et al. 2017). Nanoparticles of titanium oxide could attain the required sterilization efficacy on various bacteria such as *Vibrio anguillarum*, *Aeromonas hydrophila*, and *E. coli*, and these particles can disintegrate the organic pollution-causing agents in water (Shiwen et al. 2015). Oral administration and injection are the most dependable and potent methods of vaccination, but these methods may lead to the death of fish (Shah and Mraz 2020). Traditional methods of drug delivery in fishes performed badly due to their lower efficiency and biological presence in aquatic mediums (Sarkar et al. 2021). To avoid this problem, researchers used nanoparticles for the administration of the vaccine in fishes (Shah and Mraz 2020). Alginate particle, extracted from brown algae or as polysaccharides in a few bacteria, is a copolymer of a-L-guluronic acid and b-D-mannuronic acid, which are the primary particles responsible for vaccine delivery to fishes orally (Shah and Mraz 2020). Antibiofilm and antimicrobial properties are associated with oxides of iron, zinc, copper, and silver nanoparticles (Sarkar et al. 2021). Graphene oxide shows a hampering response to crucial aquatic disease-causing microbes such as *E. coli*, *S. aureus*, *Vibrio harveyi*, and *P. aeruginosa* in association with other polymeric nanoparticles. A small DNA strand with nano-encapsulated particles is suspended in the water body, where it is absorbed by the fishes. Subsequently, DNA is released into the fish using ultrasound, resulting in an immunological response (Sarkar et al. 2021).

Nanotechnology has a wide range of applications for water quality enhancement to facilitate a suitable environment for aquaculture. Adsorption and photocatalysis are the methods of water treatment (Shah and Mraz 2020). Shrimp aquaculture showed that the nano-enabled devices were able to enhance the quality of water and decrease the water exchange rate. It also reduces the mortality of shrimp and increases yield (Shiwen et al. 2015). The discharge of municipal sewage, industrial waste, agricultural runoff, antibiotics, etc. into water has not only caused problems for humans by reducing the amount of pure groundwater but also affected aquatic animal health. Catalytic adsorbents and hydrogel biofilms based on various nanoparticles work efficiently in water treatment (Shah and Mraz 2020). The nanoparticle can also be used in packing technology and the preservation of seafood without any microbial damage (Shiwen et al. 2015). There are different methods for packaging, like using nanopolymers and coatings to increase the strength of the packaging, which protects the fish fillet from mechanical damages or other bruising (Handy 2012). Nanomaterials such as carbon nanotubes are extremely stronger than steel, so they can be used in gear and boatmaking. Carbon nanotubes are also used in the manufacture of nets and cages (Sarkar et al. 2021).

## 4.5 Nanotechnology in Aquaculture Feed Production

The development of feed contributes the maximum input into aquaculture and growth, as well as weight gain, depending on the feed consumption by fish. A small disbalance of nutrient composition in the feed will lead to the inappropriate growth and spread of diseases, and finally, the collapse of farming occurs. Fish suffer from nutritional deficiencies because of an imbalanced diet, a lack of food, or an abundance of dietary components. In aquaculture, one of the most important functions of nanotechnology is feed production, where nanoparticles are employed for growth enhancement, vitamin delivery, and feed encapsulation (Fig. 4.1) (Pudake et al. 2019). Nanoparticles are seen to be effective for (i) growth promotion (e.g., nFe, nSe, and ZnO), (ii) delivery of micronutrients (e.g., chitosan NPs), and (iii) production of feed amount per unit time (e.g., fullerenes (C60), and nTiO2), during aquafeed development.

The growth of aquatic animals is influenced by many factors such as water quality, temperature, vaccination, and well management, but notionally balanced

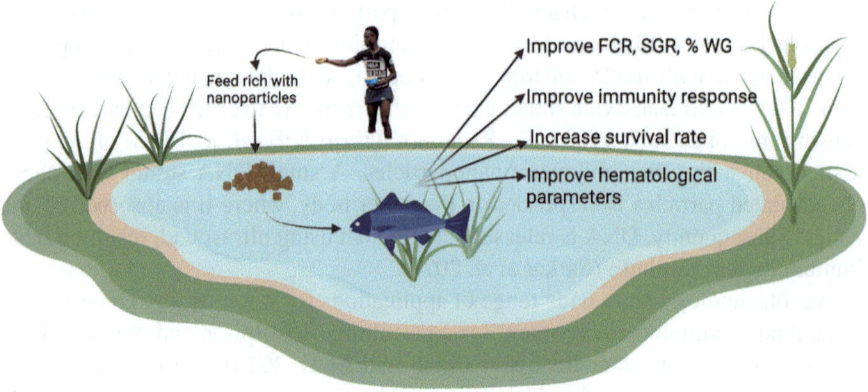


**Fig. 4.1** Demonstration of the impact of nanoparticle-rich feed on fish health. (Created with BioRender https://biorender.com/, Farmer picture by https://www.usaid.gov/. Data from Table 4.2)

aquafeed is an important factor that determines good production by improving feed digestibility, health condition, and growth performance. A well-balanced feed formulation can be achieved by using both macro- and micronutrients (Dawood et al. 2021). The use of nanotechnology/nanoparticles as feed additives is the best way to achieve proper nutrients in feed. The application of various nanoparticles as feed additives has been examined by various researchers. According to Rathore et al. 2020, selenium (Se) can boost growth hormone production, resulting in increased fish growth. It's an important trace element for animals that can be fed to them. It is a constituent of the glutathione peroxidase enzyme (GSH-Px) that upholds the cell membranes through glutathione concentration reduction (Pudake et al. 2019). The use of Se nanoparticles as a feed additive is important because of its antioxidant defence properties (Sonkusre et al. 2014). Another function of Se used as a nanoparticle is to enhance larval growth by preventing skeleton anomalies and bone mineralization (Izquierdo et al. 2017).

Silver nanoparticles (AgNPs) are also used for aquafeed development like other nanoparticles, and they have great importance for animal health due to their antibacterial activity and better efficiency due to their small size and shape and large surface-to-volume ratio that enhances the antibacterial capacity (De Silva et al. 2021). The use of AgNPs as nanoparticles increases nutrient absorption in aquatic organisms via target delivery, controlled release, encapsulation, and as cargo on a nanocarrier. AgNPs, when working as nanocarriers, indicate improved adsorption and delivery of nutraceuticals needed by an aquatic organism for better growth. The best AgNPs are obtained when they are synthesized biologically because they contain biodegradable properties, which reduce the chances of toxicity (Rather et al. 2013).

Zinc (Zn) is an important micronutrient implicated in a variety of metabolic pathways, including protein synthesis, lipid metabolism, energy consumption, and vitamin A absorption (Muralisankar et al. 2014). It has been noticed that the supplementation of zinc nanoparticles in the diet of *M. rosenbergii* postlarvae (PL) improves the performance of digestive enzymes such as protease, amylase, and lipase, as well as survival and growth (Muralisankar et al. 2014). Faiz et al. (2015) showed that zinc oxide acts as a supply of dietary Zn that demonstrates a better response of the immune system and grass carp growth improvement. A considerable increase in the concentration of antioxidant enzymes, protein content, and improved body weight of freshwater prawns (*Macrobrachium rosenbergii*) has been noticed when organisms feeding on feed improved with nZnO (Muralisankar et al. 2014).

Chitosan is indeed a polysaccharide-type nanoparticle that is frequently employed in animal feed production due to its antimicrobial capabilities, low toxicity, and low immunogenicity (Khosravi-Katuli et al. 2017). A novel application of chitosan nanoparticles is the supply of hydrosoluble or unstable micronutrients, which also increased the shelf life and liberation of vitamin C in the fish (Alishahi et al. 2014). It has been observed that *Oreochromis niloticus* shows a positive and improved growth rate and feed utilization during feeding with an improved diet with chitosan (Abd El-Naby et al. 2020). It is essential for the stimulation of phagocytic cell bactericidal activity and increased serum bactericidal activity, which results in the induction of several humoral components involved in innate and/or adaptive immunity, which successfully protect the fish from various infections (Maqsood et al. 2010). It is also used as an encapsulating agent to help release the nutrients from feed pellets into water, and it is easily degraded with contact with water (Khosravi-Katuli et al. 2017). Various nanoparticle along with the concentration used during the experiment as a feed additive for developing aquafeed has been elaborated in Table 4.2.

**Table 4.2** Effect of nanoparticles used in feed and their concentration on various fish species

| Nanoparticles | Fish species | Concentration | Effects | Reference |
|---|---|---|---|---|
| Selenium | Nile tilapia fingerlings (15.73 ± 0.05 g) | 1 mg/kg | Higher weight gain, average daily weight gain, FCR, SGR | Rathore et al. (2020) |
| | Asian seabass (Lates calcarifer) (32.78 ± 2.23 g) | 4 mg/kg | Higher feed consumption, weight gain, and specific growth<br>Enhanced immune response | Longbaf Dezfouli et al. (2019) |
| | Gilthead seabream (5.10 ± 0.43 mm) | 3 mg/kg | Prevent skeleton anomalies<br>Improve larval and growth performance | Izquierdo et al. (2017) |
| Silver | Zebra fish (Danio rerio) | 0.963 and 1.925 mg/L | Induced developmental and transcriptional distress in embryos development | Qiang et al. (2020) |

(continued)

**Table 4.2** (continued)

| Nanoparticles | Fish species | Concentration | Effects | Reference |
|---|---|---|---|---|
| Zinc | Nile tilapia (*Oreochromis niloticus*) | 60 mg/kg | High SGR rate<br>Increase in growth hormone<br>Increase in the concentration of total protein and IgM | Tawfik et al. (2017) |
| | *Macrobrachium rosenbergii* post larvae (PL) | 10–60 mg/kg | Increased growth and digestive enzyme activities (protease, amylase, and lipase)<br>Reduce the FCR | Muralisankar et al. (2014) |
| | Grass carp, (*Ctenopharyngodon Idella*) | 30 and 60 mg/kg | Showed a higher growth rate<br>Higher energy reserves available due to the higher availability of nutrients | Faiz et al. (2015) |
| | *Labeo rohita* (Hamilton) fingerlings | 20 mg/kg | Improve weight gain and SGR | Mondal et al. (2020) |
| Chitosan [poly(1,4-β-D-glucopy-ranosamine)] | *Oreochromis niloticus* (4.97 ± 0.02 g) | 5 g/kg | Enhanced the indicators related to growth performance, feed utilization<br>Improved the concentration of WBC count, RBC count, and Hb values | Abd El-Naby et al. (2020) |
| | Common carp (*Cyprinus carpio*) (45 ± 2 g) | 20.0 g/kg | Improved FCR and specific growth rate<br>Enhancement of innate immune responses<br>Reduced mortality percentage and high phagocytic activity | Maqsood et al. (2010) |
| | *Penaeus monodon* (mean initial wet weight 1.49 g) | 4.0 g/kg | Increase the survival rate<br>Improved weight gain<br>Immunostimulant | Niu et al. (2013) |

(continued)

**Table 4.2** (continued)

| Nanoparticles | Fish species | Concentration | Effects | Reference |
|---|---|---|---|---|
| Copper | White fish (*Rutilus frisii kutum*) | 10 ppm Cu | Antifungal effects of copper nanoparticles to avoid serious losses in fish hatcheries | Kalatehjari et al. (2015) |
| | Mozambique tilapia (Tilapia mossambica) | 2 mg/L and 15 mg/L | Significant improve tissues such as liver, gills, brain, and body weight<br>Modify the enzyme performance in various organs, e.g., liver, gills, brain, etc.<br>Overall reduction in protein level | Al Ghais et al. (2019) |
| Titanium | Rainbow trout (*Oncorhynchus mykiss*) | 10 and 100 mg/kg | Significantly body weight gain, but mean final weight approximately same for all group<br>No major haematological disturbances | Ramsden et al. (2009) |
| | Common carp (Cyprinus carpio) (15.18 ± 0.22 g) | 0.125 mg/L | No significant effect on blood parameters such as haemoglobin concentrations<br>The immune parameters of serum and mucus significantly decrease | Hajirezaee et al. (2020) |

(continued)

**Table 4.2** (continued)

| Nanoparticles | Fish species | Concentration | Effects | Reference |
|---|---|---|---|---|
| Iron | Goldfish (*Carassius auratus*) (4.3 g) | 0.5 g/kg | Significantly improved feed conversion ratio and specific growth rates<br>Increase the concentrations of alkaline phosphatase in plasma<br>Upregulated IGF-1 and ghrelin gene expressions | Akbary and Jahanbakhshi (2019) |
| | *Labeo rohita* | 10.0 mg/kg | Improve the weight gain, survival rate, SGR, and FCR<br>Increase the concentration of carbohydrates in the liver, gills, muscles of the fish | Thangapandiyan et al. (2020) |
| | Bagridae catfish (*Clarias batrachus*) (5.23 ± 0.07 g) | 40 mg/kg | Highest growth and feed utilization performance has been observed<br>Significantly increase the total protein and lipid content of fish muscle<br>Serum total protein, cholesterol, and triglyceride were found to increase | Akter et al. (2018) |

Feed formulation along with nanoparticles can be done by various methods. Faiz et al. (2015) formed the zinc nanoparticle-rich feed. The inorganic dietary sources of Zn, $ZnSO_4$, and ZnO were used to synthesize zinc oxide nanoparticles (ZnO-NP), and it is ground along with other feed ingredients (dry) to obtain a fine powder that is mixed with oil. The semisolid paste is prepared by adding water in the optimal quantity and passing it through a meat grinder to obtain an unbroken noodle-like shape that changes into small, uniform-size pellets. At low light and room temperature, the pellets were dried to avoid oxidation and stored in an airtight container at 4 °C. It is important to note that the manufacturing of zinc nanoparticle-rich feed is done by grinding ingredients as per respective diets and mixing with inorganic sources of Zn after and before being blended with oil (Faiz et al. 2015). Thangapandiyan et al. (2020) used the leaves of *Amaranthus tricolor* for the green synthesis of iron oxide nanoparticles and mixed them with ingredients for achieving the desired feed diet. Ramsden et al. (2009) formulate a diet rich in titanium dioxide. One kilogram of feed was placed in the commercial feed mixer and sprayed with the required $TiO_2$ nanoparticle solution. The spraying of a 10% bovine gelatine solution on ingredients leads to the $TiO_2$ nanoparticles being rapidly covered and sealing the feed. The gelatin coating was permitted to dry before the feed was moved to airtight storage containers (Ramsden et al. 2009). Maqsood et al. (2010) ground all desired ingredients separately in an electric grinder and mixed them properly, and water was added in the optimal quantity. The well-mixed mixture of ingredients was steamed for 20–25 minutes. The chitosan along with the mineral mixture at the required quantities was added and mixed thoroughly for making dough. The dough of the desired consistency passed through a hand pelletizer with the desired size of the die, and the resultant pellets were put under shade for air-drying. The dried pellets were stored at a temperature of −20 °C after being packed in airtight polythene. After achieving the pellets along with the desired nanoparticles, they can be used for feeding the aquatic organism.

## 4.6 Nanoparticle Exposure to Aquatic Animals and Health Improvement

The field of nanotechnology is still growing in aquaculture. Recent observations proved that the role of nanoparticles (NPs) in the aquaculture sector is increasing gradually. Various companies related to nanotechnology tools are endeavouring the use of nanotechnology-based tools to eradicate the obstacles around aquatic food, reproduction, species culture, growth, health, and treatment of water to upsurge the rates of overall aquaculture production (Khosravi-Katuli et al. 2017).

In the fishery and its related industries, NPs are developed for various direct and indirect purposes, as shown in Fig. 4.2. Fishpond sterilization, treatment of wastewater, and fish processing for commercialization such as tagging and barcoding are some of the indirect uses; direct uses include animal health management like controlling fish disease and the feed industry (Khosravi-Katuli et al. 2017).

### *4.6.1 Silver Nanoparticles*

Silver nanoparticles effectively control several diseases caused by viruses, fungi, bacteria, and other single-cellular microorganisms. It holds back the reproduction of various disease-causing microorganisms. Reproduction and growth of those microorganisms responsible for the contagion are inhibited by bad odour, inflammation, and blisters. AgNPs are noticed to act quickly and to be nontoxic, freshening, highly efficient, non-stimulating, tolerance-free, nonallergic, and hydrophilic, thus making them very efficient for bacterial resistance. Hence, AgNPs act as disinfectants in aquaculture to disinfect and thwart diseases (Deshmukh et al. 2019).

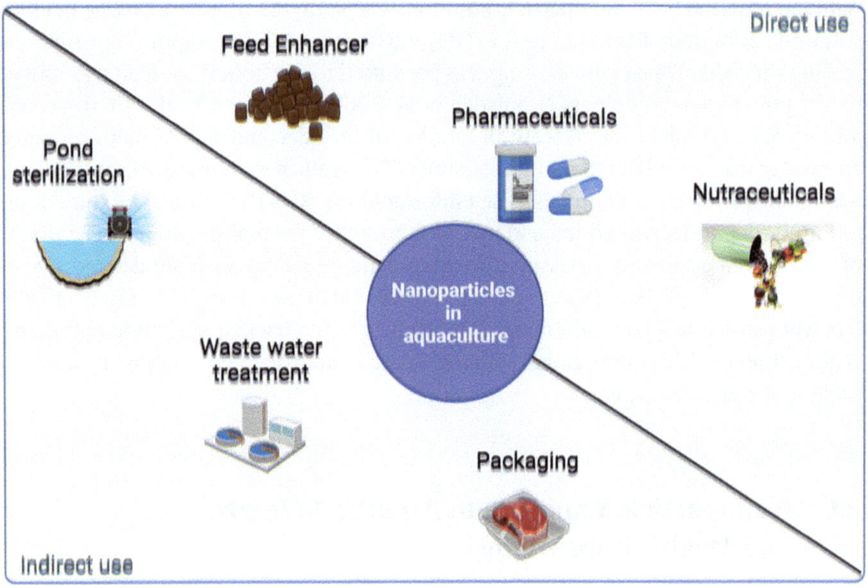


**Fig. 4.2** Uses of nanoparticles in the aquaculture sector. (Data from Khosravi-Katuli et al. (2017); created with BioRender.com)

Various researchers investigated the antibacterial mechanisms of AgNPs. Silver can interact with the membranes of a bacterial cell, which comprise proteins and amino acids which contain sulphur and are found on both the inside and outside of the cell membrane, resulting in the inactivation of the bacterial cell. Furthermore, silver ions released from AgNPs inhibit enzyme activities when interacting with sulphur-containing proteins as well as with phosphorus in DNA. AgNPs engage in various roles such as antimicrobial and anti-adhesive agents to restrain bacterial adsorption, growth, and attachment on the surface of the membrane, resulting in the prevention of biofilm formation on the membrane (Deshmukh et al. 2019). Banach et al. in 2016 reported that AgNPs are efficient in obliterating a wide range of Gram-positive (+) and Gram-negative (−) bacteria. Gram-positive bacteria consist of genera such as *Clostridium*, *Enterococcus*, *Bacillus*, *Listeria*, *Streptococcus*, and *Staphylococcus*, while Gram-negative (−) bacteria consist of the genera *Pseudomonas*, *Acinetobacter*, *Vibrio*, *Salmonella*, and *Escherichia* (Banach et al. 2016).

### *4.6.2 Chitosan Nanoparticle*

Aquatic organisms exposed to pesticide pollutants demonstrate an inequity between cell-produced reactive oxygen species (ROS) endogenously as well as exogenously, which can then encourage a decline in antioxidant defence systems or instigate oxidative impairment in the cells of organisms. Enzymatic and nonenzymatic antioxidants are crucial for preventing the destruction of the cellular membrane by ROS and re-establishing normal function and cellular metabolism. Vitamin C, or ascorbic acid, is one of the chief antioxidants, nonenzymatic in nature that can transform ROS to stop their destructive effects, thus shielding tissue from impairment (Naiel et al. 2020). Özkan et al. 2012 mentioned that vitamin C can preserve both cytosolic cells and membrane elements from oxidative impairment. Numerous researchers have stated that diets enhanced with vitamin C vetoed the bad effects of stress for aquatic animals, decreased the destructive effects of water toxicity contamination, and boosted immune defence mechanisms (Naiel et al. 2020).

A natural deacetylation polymer, chitosan, is changed from chitin and separated mostly from the crustacean exoskeleton. Chitosan and its reforms have been fruitfully used for the purging of metal phenols, dyes, ions, pesticides, fungicides, and humic constituents through adsorption (Naiel et al. 2020). Chitosan has an advantage over activated carbon and other sorbents due to its low cost, a prominent affinity for numerous toxins (because of the existence of amino and hydroxyl groups), high reactivity, chemical constancy, and selectivity concerning pollution. Besides the above-mentioned reasons, chitosan-supplemented diets with nanoparticle interpretation play a critical role in the stoppage of detrimental toxicological outcomes of pesticides in aquatic environments (Naiel et al. 2020).

### 4.6.3 Selenium (Se)

Selenium (Se) is an imperative microelement, necessary for the healthier performance and well-growth of aquatic animals by enhancing immunity response and tolerance of aquatic species to infectious diseases (Dawood et al. 2019, 2021). Adequate Se add-on is essential for various metabolic processes, such as the production of thyroid hormone, fecundity, DNA synthesis, and cytokine formation (Dawood et al. 2021). Supplementation of Se in fish feed was widely conventional since Se from feed and ambient water alone could not provide the optimal level needed by cultured aquatic species (Dawood et al. 2019). They act as a precursor for synthesizing antioxidative enzymes, resulting in high antioxidative ability. Se insufficiency reduced appetite and growth as well as root peroxidative destruction to cell membranes, resulting in elevated mortality in rare cases, while extreme Se lessened the efficiency of feeding and brought about tissue destruction, reproductive failure, and teratogenic abnormalities of organs (e.g., mouth, spine, etc.) (Watanabe et al. 1997).

Therefore, an ideal dietary level of Se is required for diverse cultures of various fish species. Selenium can shield the oxidation damage of animal cells caused by various stressors due to poor water quality, high density, transportation, and infectious diseases by fostering the activity of thyroid hormone metabolism and reproduction and antioxidant-related enzymes (Pacitti et al. 2016; Hefnawy and Tórtora-Pérez 2010). The aquatic organisms demand more precise diets which consist of the micro and macro ingredients to meet their high rate of metabolism coupled with their improved growth rate. The supplementation of Se in fish feed can be influenced by Se structure, feed formulation, and the species and size of fish (Dawood et al. 2019). Recently, the Se nanoparticle (Se-NP) form was used for fish diet due to its low harmfulness and high bioavailability. Additionally, nano-Se enhanced the growth of the animal, feed utilization, and antioxidant defence ability of several cultured fish (Dawood et al. 2019).

### *4.6.4 Zinc (Zn)*

Fish require dietary zinc for their healthy growth and development. It acts as an antimicrobial and regulates both the reproductive and immune systems of animals, in addition to promoting growth. As Zn cannot be deposited in the bodies of animals, regular dietary intake is a must. Failure of this could lead to repeated infections and a poor appetite (Chris et al. 2018).

### *4.6.5 Iron (Fe)*

The role of iron in physiological processes includes transporting oxygen molecules, lipid oxidation reactions, cell respiration, defence against infections, and immune system functioning. Iron portrays a vital role in aquatic species. Most of the natural iron sources do not meet the dietary requirements of fish even if they are found abundant due to their low bioavailability and solubility; thus, dietary iron supplementation became essential. Significant progress in the growth, survivability, and hematological parameters of fish is detected when nurtured with a nano-iron-supplemented diet (Chris et al. 2018).

## 4.7 Future Perspectives

Nanotechnology is still evolving and paving the way in aquaculture with its very few applications in feed enhancement, pharmaceuticals, and nutraceuticals. Nanoparticles, in general, are used to improve vitality, immune mechanism, growth, fecundity, reproductive ability, cell defence, etc. as they hold various properties such as antimicrobial, antibacterial, antioxidant, and disinfectant. As the chapter focuses on the interpretation of nanotechnology in feed development, only a few researches are successful in this area due to many disadvantages. Each culture species requires a specific feed ameliorated with an ideal concentration of nanoparticles, and this was successfully derived theoretically. Practical implementation has made it difficult to produce a feed that is ideal for a wide range of aquatic organisms in various water conditions with different salinity ranges. Hence, the field of nanotechnology still requires a lot of expertise to develop advanced products.

## References


Abad-Álvaro I, Trujillo C, Bolea E, Laborda F, Fondevila M, Latorre MA, Castillo JR (2019) Silver nanoparticles-clays nanocomposites as feed additives: characterization of silver species released during in vitro digestions. Effects on silver retention in pigs. *Microchem J* 149:104040

Abd El-Naby AS, Al-Sagheer AA, Negm SS, Naiel MA (2020) Dietary combination of chitosan nanoparticle and thymol affects feed utilization, digestive enzymes, antioxidant status, and intestinal morphology of Oreochromis niloticus. Aquaculture 515:734577

Ahmed F, Soliman FM, Adly MA, Soliman HA, El-Matbouli M, Saleh M (2019) Recent progress in biomedical applications of chitosan and its nanocomposites in aquaculture: a review. Res Vet Sci 126:68–82

Akbary P, Jahanbakhshi A (2019) Nano and macro iron oxide (Fe2O3) as feed additives: effects on growth, biochemical, activity of hepatic enzymes, liver histopathology and appetite-related gene transcript in goldfish (Carassius auratus). Aquaculture 510:191–197

Aklakur M, Rather MA, Kumar N (2016) Nanodelivery: an emerging avenue for nutraceuticals and drug delivery. J Crit Rev Food Sci Nutr 56:2352–2361

Akter N, Alam MJ, Jewel MAS, Ayenuddin M, Haque SK, Akter S (2018) Evaluation of dietary metallic iron nanoparticles as feed additive for growth and physiology of Bagridae catfish Clarias batrachus (Linnaeus, 1758). Int J Fish Aquat Stud 6(3):371–377

Al Ghais S, Bhardwaj V, Kumbhar P, Al Shehhi O (2019) Effect of copper nanoparticles and organometallic compounds (dibutyltin) on tilapia fish. J Basic Appl Zool 80(1):32

Alishahi A, Proulx J, Aider M (2014) Chitosan as biobased nanocomposite in seafood industry and aquaculture. In: Seafood science: advances in chemistry, technology and applications, vol 211. CRC Press, Boca Raton

Baby R, Saifullah B, Hussein MZ (2019) Carbon nanomaterials for the treatment of heavy metal-contaminated water and environmental remediation. Nanoscale Res Lett 14(1):1–17

Baldissera MD, Souza CF, Alessio KO, Krawczak KW, Abbad LB, da Silva AS, Bizzi C, Ourique AF, Zeppenfeld CC, Baldisserotto B, Cunha MA (2020) Diphenyl diselenide-loaded nanocapsules in silver catfish feed enhance growth, improve muscle antioxidant/oxidant status and increase selenium deposition: advantages of nanotechnology for fish health. Aquacult Res 51:1–3

Banach M, Tymczyna L, Chmielowiec-Korzeniowska A, Pulit-Prociak J (2016) Nanosilver biocidal properties and their application in disinfection of hatchers in poultry processing plants. Bioinorg Chem Appl 2016:1

Bradder P, Ling SK, Wang S, Liu S (2011) Dye adsorption on layered graphite oxide. J Chem Eng Data 56(1):138–141

Bunglavan SJ, Garg AK, Dass RS, Shrivastava S (2014) Use of nanoparticles as feed additives to improve digestion and absorption in livestock, 38

Chaudhry Q, Castle L (2011) Food application of nanotechnologies: an overview of opportunities and challenges for developing countries. Trends Food Sci Technol 22:599

Chen H, Weiss J, Shahidi F (2006) Nanotechnology in nutraceuticals and functional foods. Food Technol (Chicago) 60(3):30–36

Chris UO, Singh NB, Agarwal A (2018) Nanoparticles as feed supplement on growth behaviour of cultured catfish (Clarias gariepinus) fingerlings. Mater Today Proc 5(3):9076–9081

Dawood MA, Koshio S, Zaineldin AI, Van Doan H, Ahmed HA, Elsabagh M, Abdel-Daim MM (2019) An evaluation of dietary selenium nanoparticles for red sea bream (Pagrus major) aquaculture: growth, tissue bioaccumulation, and antioxidative responses. Environ Sci Pollut Res 26(30):30876–30884

Dawood MA, Basuini MFE, Yilmaz S, Abdel-Latif HM, Kari ZA, Abdul Razab MKA et al (2021) Selenium nanoparticles as a natural antioxidant and metabolic regulator in aquaculture: a review. Antioxidants 10(9):1364

De Silva C, Nawawi NM, Abd Karim MM, Abd Gani S, Masarudin MJ, Gunasekaran B, Ahmad SA (2021) The mechanistic action of biosynthesised silver nanoparticles and its application in aquaculture and livestock industries. Animals 11(7):2097

Deng Y, Cheng Q (2003) Affects of nano-selenium on the growth of Nile tilapia (Oreochromis niloticus). Inland Aquat Prod 6:28–30

Deshmukh SP, Patil SM, Mullani SB, Delekar SD (2019) Silver nanoparticles as an effective disinfectant: a review. Mater Sci Eng C 97:954–965

Faiz H, Zuberi A, Nazir S, Rauf M, Younus N (2015) Zinc oxide, zinc sulfate and zinc oxide nanoparticles as source of dietary zinc: comparative effects on growth and hematological indices of juvenile grass carp (Ctenopharyngodon idella). Int J Agric Biol 17(3):568

FAO (2020) The state of world fisheries and aquaculture. Food and Agriculture Organization of the United Nations, Rome

Fathi M, Mozafari MR, Mohebbi M (2012) Nanoencapsulation of food ingredients using lipid based delivery systems. Trends Food Sci Technol 23:13–27

FSAI (Food Safety Authority of Ireland) (2008) The relevance for food safety of applications of nanotechnology in the Food and Feed Industries Abbey Court, Lower Abbey Street, Dublin

Gangadoo S, Stanley D, Hughes RJ, Moore RJ, Chapman J (2016) Nanoparticles in feed: progress and prospects in poultry research. Trends Food Sci Technol 58:115–126

Hajirezaee S, Mohammadi G, Naserabad SS (2020) The protective effects of vitamin C on common carp (Cyprinus carpio) exposed to titanium oxide nanoparticles (TiO2-NPs). Aquaculture 518:734734

Handy RD (2012) FSBI briefing paper: nanotechnology in fisheries and aquaculture. Fisheries Society of the British Isles, 4, 1–29

Izquierdo MS, Ghrab W, Roo J, Hamre K, Hernández-Cruz CM, Bernardini G et al (2017) Organic, inorganic and nanoparticles of Se, Zn and Mn in early weaning diets for gilthead seabream (Sparus aurata; Linnaeus, 1758). Aquac Res 48(6):2852–2867

Jimenez-Fernández E, Ruyra A, Roher N, Zuast E, Infante C, Fernández-Díaz C (2014) Nanoparticles as a novel delivery system for vitamin C administration in aquaculture. Aquaculture 432:426–433

Kalatehjari P, Yousefian M, Khalilzadeh MA (2015) Assessment of antifungal effects of copper nanoparticles on the growth of the fungus Saprolegnia sp. on white fish (Rutilus frisii kutum) eggs. Egypt J Aquat Res 41(4):303–306

Khan FA, Khan F (2020) Applications of nanomaterials in human health. Springer, Singapore

Khosravi-Katuli K, Prato E, Lofrano G, Guida M, Vale G, Libralato G (2017) Effects of nanoparticles in species of aquaculture interest. Environ Sci Pollut Res 24(21):17326–17346

Longbaf Dezfouli M, Ghaedtaheri A, Keyvanshokooh S, Salati AP, Mousavi SM, Pasha-Zanoosi H (2019) Combined or individual effects of dietary magnesium and selenium nanoparticles on growth performance, immunity, blood biochemistry and antioxidant status of Asian seabass (Lates calcarifer) reared in freshwater. Aquac Nutr 25(6):1422–1430

Louros VL, Ferreira LM, Silva VG, Silva CP, Martins MA, Otero M, Esteves VI, Lima DL (2021) Photodegradation of aquaculture antibiotics using carbon dots-TiO2 nanocomposites. *Toxics* 9(12):330

Majhi KC, Yadav M (2021) Synthesis of inorganic nanomaterials using carbohydrates. In: Green sustainable process for chemical and environmental engineering and science. Elsevier, Amsterdam, pp 109–135

Maqsood S, Singh P, Samoon MH, Khansaheb Balange A (2010) Effect of dietary chitosan on non-specific immune response and growth of Cyprinus carpio challenged with Aeromonas hydrophila. Int Aquat Res 2(2):77–85

Márquez JCM, Partida AH, del Carmen M, Dosta M, Mejía JC, Martínez JAB (2018) Silver nanoparticles applications (AgNPS) in aquaculture. Int J Fish Aquat Stud 6(2):5–11

Moges FD, Paritosh P, Parashar SKS, Das B (2020) Mechanistic insights into diverse nano-based strategies for aquaculture enhancement: a holistic review. Aquaculture 519:734770, 2

Mondal AH, Behera T, Swain P, Das R, Sahoo SN, Mishra SS, Das J, Ghosh K (2020) Nano zinc vis-à-vis inorganic Zinc as feed additives: effects on growth, activity of hepatic enzymes and non-specific immunity in rohu, Labeo rohita (Hamilton) fingerlings. Aquac Nutr 26(4):1211–1222

Munawar N (2021) Interaction and applications of nanoparticles in fishes and aquaculture. Sch Bull 7(6):150–155

Muralisankar T, Bhavan PS, Radhakrishnan S, Seenivasan C, Manickam N, Srinivasan V (2014) Dietary supplementation of zinc nanoparticles and its influence on biology, physiology and immune responses of the freshwater prawn, Macrobrachium rosenbergii. Biol Trace Elem Res 160(1):56–66

Naiel MA, Ismael NE, Abd El-hameed SA, Amer MS (2020) The antioxidative and immunity roles of chitosan nanoparticle and vitamin C-supplemented diets against imidacloprid toxicity on Oreochromis niloticus. Aquaculture 523:735219

Niu J, Lin HZ, Jiang SG, Chen X, Wu KC, Liu YJ, Wang S, Tian LX (2013) Comparison of effect of chitin, chitosan, chitosan oligosaccharide and N-acetyl-d-glucosamine on growth performance, antioxidant defenses and oxidative stress status of Penaeus monodon. Aquaculture 372:1–8

Özkan F, Gündüz SG, Berköz M, Hunt AÖ, Yalın S (2012) The protective role of ascorbic acid (vitamin C) against chlorpyrifos-induced oxidative stress in Oreochromis niloticus. Fish Physiol Biochem 38:635–643

Pacitti D, Lawan MM, Feldmann J, Sweetman J, Wang T, Martin SAM, Secombes CJ (2016) Impact of selenium supplementation on fish antiviral responses: a whole transcriptomic analysis in rainbow trout (Oncorhynchus mykiss) fed supranutritional levels of Sel-Plex®. BMC Genomics 17(1):1–26

Peters R, ten Dam G, Bouwmeester H, Helsper H, Allmaier G, vd Kammer F, Ramsch R, Solans C, Tomaniová M, Hajslova J, Weigel S (2011) Identification and characterization of organic nanoparticles in food. TrAC Trends Anal Chem 30(1):100–112

Peters RJ, Bouwmeester H, Gottardo S, Amenta V, Arena M, Brandhoff P, Marvin HJ, Mech A, Moniz FB, Pesudo LQ, Rauscher H (2016) Nanomaterials for products and application in agriculture, feed and food. Trends Food Sci Technol 54:155–164

Pudake RN, Chauhan N, Kole C (eds) (2019) Nanoscience for sustainable agriculture. Springer International Publishing, Cham, p 711

Qiang L, Arabeyyat ZH, Xin Q, Paunov VN, Dale IJ, Lloyd Mills RI, Rotchell JM, Cheng J (2020) Silver nanoparticles in zebrafish (Danio rerio) embryos: uptake, growth and molecular responses. *Int Journal of Molecular Sciences 21*(5):1876

Ramsden CS, Smith TJ, Shaw BJ, Handy RD (2009) Dietary exposure to titanium dioxide nanoparticles in rainbow trout, (Oncorhynchus mykiss): no effect on growth, but subtle biochemical disturbances in the brain. Ecotoxicology 18(7):939–951

Rather MA, Sharma R, Aklakur M, Ahmad S, Kumar N, Khan M, Ramya VL (2011) Nanotechnology: a novel tool for aquaculture and fisheries development. A prospective mini-review. Fish Aquacult J 3:16

Rather MA, Sharma R, Gupta S, Ferosekhan S, Ramya VL, Jadhao SB (2013) Chitosan-nanoconjugated hormone nanoparticles for sustained surge of gonadotropins and enhanced reproductive output in female fish. PLoS One 8(2):e57094

Rathore SS, Murthy HS, Abdullah-Al Mamun M, Nasren S, Rakesh K, Kumar BTN, et al. (2020) Nano-selenium supplementation to ameliorate nutrition physiology, immune response, antioxidant system and disease resistance against Aeromonas hydrophila in Monosex Nile Tilapia (Oreochromis niloticus). Biol Trace Elem Res 99(8):3073–3088. https://doi.org/10.1007/s12011-020-02416-0. PMID: 33025518.

Sarkar B, Mahanty A, Gupta SK, Choudhury AR, Daware A, Bhattacharjee S (2021) Nanotechnology: a next-generation tool for sustainable aquaculture. Aquaculture 546:737330

Sekhon BS (2014) Nanotechnology in agri-food production: an overview. Nanotechnol Sci Appl 7:44
Shaalan M, Saleh M, El-Mahdy M, El-Matbouli M (2016) Recent progress in applications of nanoparticles in fish medicine: a review. Nanomedicine 12(3):701–710
Shah BR, Mraz J (2020) Advances in nanotechnology for sustainable aquaculture and fisheries. Rev Aquac 12(2):925–942
Shiwen H, Ling W, Lianmeng L, Yuxuan H, Lu L (2015) Nanotechnology in agriculture, livestock, and aquaculture in China. A review. Agron Sustain Dev 35:369–400
Sonkusre P, Nanduri R, Gupta P, Cameotra SS (2014) Improved extraction of intracellular biogenic selenium nanoparticles and their specificity for cancer chemoprevention. J Nanomed Nanotechnol 5(2):1
Tawfik M, Moustafa M, Abumourad IMK, El-Meliegy E, Refai M (2017, August). Evaluation of Nano zinc oxide feed additive on tilapia growth and immunity. In 15th international conference on environmental science and technology, Rhodes, Greece (Vol. 1342, No. 1, pp. 1–9)
Thangapandiyan S, Alisha AA, Anidha K (2020) Growth performance, hematological and biochemical effects of iron oxide nanoparticles in Labeo rohita. Biocatal Agric Biotechnol 25:101582
Tórtora-Pérez JL (2010) The importance of selenium and the effects of its deficiency in animal health. Small Rumin Res 89(2–3):185–192
Watanabe T, Kiron V, Satoh S (1997) Trace minerals in fish nutrition. Aquaculture 151(1–4):185–207
Zhou X, Wang Y, Gu Q, Li W (2009) Effects of different dietary selenium sources (selenium nanoparticle and selenomethionine) on growth performance, muscle composition and glutathione peroxidase enzyme activity of crucian carp (Carassius auratus gibelio). Aquaculture 291(1–2):78–81

# Chapter 5
# Future Prospects of Nanotechnology in Aquaculture

**A. Sai Ramesh, N. V. P. Meghana, Ayurika Tripathy, L. Karthik, G. Kiran, and Vishnu Kirthi Arivarasan**

## 5.1 Introduction

### *5.1.1 Intervention of Nanotechnology and Aquaculture*

Nanotechnology focuses on the ability to manipulate matter at the molecular and atomic levels, and as a result, it has enormous potential for developing new materials with improved qualities (Sarkar et al. 2015). Nanoparticles (NPs) play an important role in nanotechnology; new NPs have stimulated the growth of nanoscale materials and the rapid expansion of their applications in numerous fields. Because of their small size and high surface-to-volume ratio, NPs are important advertisements in a few initiatives and the development of research areas (Shah and Mraz 2020). Aquaculture is the world's fastest-growing food-producing sector, and it essentially contributes to the world's fish stockpile for human consumption (Diallo 2011). To ensure a sustainable development that meets global needs, the aquaculture movement must overcome a number of negative aspects arising from its own practices, such as the high concentration of natural combinations in untreated wastewater, the extensive use of anti-toxins, and the proliferation of disease vectors.

A. S. Ramesh (✉) · N. V. P. Meghana · A. Tripathy
Department of Biotechnology, Vel Tech High Tech Dr. Rangarajan Dr. Sakunthala Engineering College, Chennai, Tamil Nadu, India

L. Karthik
Synthetic Biology, Research & Department, Salem Microbes Pvt Ltd, Salem, Tamil Nadu, India

G. Kiran
School of Pharmacy, Anurag University, Hyderabad, Telangana, India

V. K. Arivarasan (✉)
Department of Microbiology, Lovely Professional University, Phagwara, Punjab, India



V. K. Arivarasan et al. (eds.), *Nanotechnological Approaches to the Advancement of Innovations in Aquaculture*, Nanotechnology in the Life Sciences, https://doi.org/10.1007/978-3-031-15519-2_5

Nanotechnology could provide a variety of solutions to such problems, ensuring a manageable increase in the aquaculture movement.

## 5.2 Future Prospects of Nanotechnology in Aquaculture

Intervention of nanotechnology and aquaculture has come a long way since its beginning. Numerous concepts of nanotechnology have proven effective in increasing the efficiency of aquaculture. A few of the many are as follows:

- Nanoparticles in water purification to reduce toxicity.
- Nanosensors in fish pond cleaning to reduce the probability of contamination in the pond, which may lead to eutrophication.
- Nanoparticles in nutrient distribution.
- Nanotechnology-based vaccines which play a vital role in aquaculture as a defense mechanism against viruses and protecting the host organisms.
- While the growing of aquatic animals and plants may seem to be a relatively new trend, aquaculture actually tackles a nearly 4000-year-old activity. An Egyptian burial site from before 2000 BC is engraved with *Tilapia nilotica* lake culture. Furthermore, carp farming may have provided a source of new fish for the Chinese monarch circa 1000 BC. This erroneous perception may be due to the fact that aquaculture production, as well as its use in human consumption, has increased dramatically in the last three decades.
- Various advantages in receipt of the aquaculture movement.
- Creation of great food, increment of family food supply, and improvement of sustenance.
- Fortified minimal economies by the increment in business and decrease of food prices.
- Improvement of water assets and supplement the executives at family or then again local area levels.
- Conservation of amphibian biodiversity through re-loading and recuperating of ensured species.
- Whenever done reasonably, the board and protection of wild stock pushed by business and sports fisheries.
- Excitement of exploration and innovation improvement.
- Expanded schooling and natural mindfulness.

With a global population of about 7 billion people, the demand for ocean food is only going to grow, necessitating additional expansion and expansion of aquaculture production. To achieve sustainable growth that satisfies global needs, the aquaculture movement must overcome a few unfavorable viewpoints arising from its own training, which could have bad effects on the initial creation and have severe environmental implications (Fig. 5.1).

Nanotechnology is a developing innovation with high potential for application in the aquaculture business. Nanoengineered minerals (<100 nm) are widely perceived

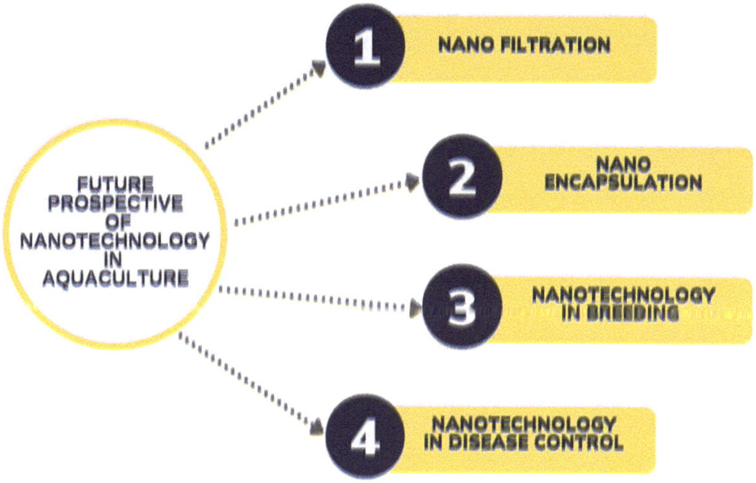


**Fig. 5.1** Schematic representation of various future prospects of nanotechnology in aquaculture

in aqua feed due to various factors such as their high dissolvability and dynamic surfaces.

Nanominerals are described by the following factors:

- Higher surface region partiality
- Higher solvency
- Warm opposition
- Low harmfulness
- Slow discharge rate
- Supported delivery

The selection of nanomaterial is based on natural capacity, physiological condition and its metabolic properties.

### *5.2.1 Nanofiltration*

For improved liberation of new components from lake water, nanofiltration technologies are suggested over, for the most part, micron-size filtering. Nanocomposite layers containing nanosilver and polyamide (PA) have been shown to have antimicrobial and biofouling sway on *Pseudomonas* sp. in the presence of water alteration and salt excusal sway (Zodrow et al. 2009). As a result, nanosilver composite can be used to create multifunctional films that act as filtration systems for separating dirty water. Using dendritic polymers, a monodispersive and considerably spread nanosize material, achievements in macromolecular science open up new avenues for designing essential ultrafiltration technologies for water purification. Dendrite polymers can be used to purify lake water because of their carefully designed and planned structure (Zahid et al. 2018; Joseph et al. 2019).

## 5.2.2 *Nanoencapsulation*

Nanoencapsulation protects food's delicate and vital bioactive components against a variety of harsh environmental conditions (Fathi et al. 2012). They were involved in the eradication of contradictions and the concealment of objectionable odors and tastes through solubilization, as well as assisting in the revealing of taste. The addition of single-walled carbon nanotubes (SWCNTs) to trout feed, which occurs during the manufacture of hard feed pellets, is an example of nanoencapsulation innovation (Meliana et al. 2017; Aklakur et al. 2016).

### 5.2.2.1 Selenium Nanoparticles

The role of nanotechnology in aquaculture could be explained using the example of selenium particles. Selenium (Se) particles are among the microelements involved in a variety of activities in a wide range of aquatic animals. The primary function of selenium is to protect cells from oxidation by developing anti-oxidative defenses. Se also functions as a cofactor in the assembly of selenoproteins involved in the catalysis of hydrogen peroxide (Karamzadeh et al. 2021). Se also aids in the digestion of the thyroid organ, as well as the multiplication and improvement of body tissues. Se possesses natural and nonnatural structures that have low bioavailability, dissolvability, and adhesion qualities at first. Accordingly, presenting Se nanoparticles in the aqua feed business is enthusiastically prescribed to augment sea-going creatures' well-being status and usefulness. Se nanoparticles are extensively researched due to their high potential as a development promoter, an antioxidant, and an immunostimulant specialist in aquaculture. Different examinations revealed the need to include Se nanoparticles for upgrading the development execution, physiological, and well-being status in oceanic creatures (Deilamy Pour et al. 2021). Because of the low edge between the advantages and the poisonousness, Se nanoparticles must be remembered for aqua feed in light of a portion explicit way. Unnecessary degrees of dietary Se are unsafe for most creatures, including domesticated animals and fish (Nuttall 2006). The clever Se nanoparticles are characterized by their minimal toxicity and tremendous use. The use of Se nanoparticles, in particular, improved the development, implementation, and efficiency of sea animals. The different roles of Se nanoparticles are presented and explored in this research, with a focus on development progress and feed use impacts, metabolic guideline roles, anti-oxidative and physiological views, and the role of Se nanoparticles as an antistress specialist (Mahdave Jehanabad et al. 2019; Kumar and Singh 2019).

### 5.2.3 *Fish Breeding*

Aquaculture research includes a lot of work on fish reproduction and rearing. The most significant advancement is the multiphase transit of compounds by infusions or supplements for the purpose of sustaining a brood stock, which is done to build gonadal tissue. In this manner, animating substances such as human chorionic gonadotropin (HCG) and others are delivered from the prespawning stage, and fish are confronted with pressure, word-related anguish, and other issues. Regular intake of maturational substances such as progesterone or testosterone adds to the problem because they drain into the surrounding water during transportation (Kailasam et al. 2006; Kumari et al. 2013). To dispose of these issues, the implantation of hormonal pellets under the skin was endeavored. Nano-epitomized hormonal conveyance was found to be a more viable option in contrast to this methodology. Chemically stacked nanocarriers can be embedded at the pre-generating stage under the fish skin to trigger development because of their slow delivery and also high maintenance time (Toranzo et al. 2005; Verma et al. 2018).

### 5.2.4 *Fish Disease Control*

Infection is one of the most serious threats to the concentrated aquaculture system, and the board of health is critical for preventing pathogenic invasion and recovering fish species. Nanotechnology, like regular innovation, can make a significant contribution in these areas by employing novel tactics (Fig. 5.2).

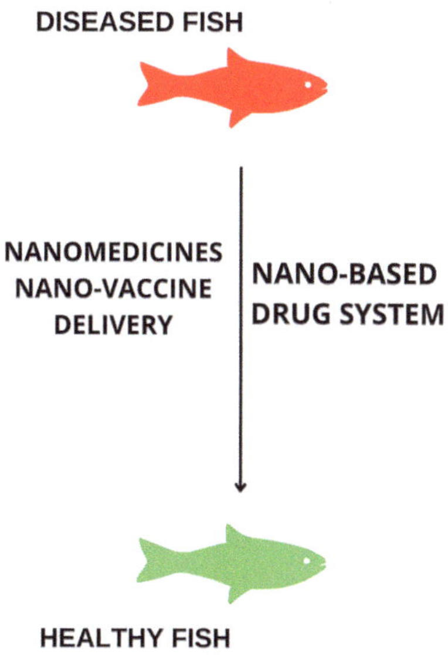


**Fig. 5.2** Schematic representation of using nanotechnology-based drugs in aquaculture

Nanomedicine is a fast-emerging field of nanotechnology that offers a wealth of opportunities to improve fish health by utilizing the intrinsic properties of many types of nanoparticles. Nanomaterials with antimicrobial and preventive capabilities, such as nanosilver and zinc oxide nanoparticles, are already being used to lower pathogenic loads in aquaculture systems.

This one-of-a-kind nanomedicine phenomenon is ubiquitous, nonspecific, and extensively applicable. Nanoparticles such as titanium dioxide and copper oxide have antibacterial properties and could be useful nanomedicines for fish.

### *5.2.5 Nanosensors*

Another nanotechnology-mediated method that is used to enhance aquaculture is by using "nanosensors." Nanosensors to screen the stock, cleaning fish lakes with nanotechnology gadgets, and DNA nanovaccines in nanocases containing short strands of DNA are added to a fish lake, where they are consumed by the cells of the fish. Ultrasound is then used to burst the cases, delivering the DNA and getting a safe reaction from the fish. Utilizing iron nanoparticles to accelerate the development of fish and a robotized drug conveyance framework is extremely quick to become reality with this imaginative innovation (Sethi and Panigrahi 2011).

### *5.2.6 Nanoparticles in Effluent Treatment*

Nanoparticles are acquiring ubiquity for their applications in the evacuation of microorganisms, organics, inorganic synthetics, halogenated intensifiers like pesticides, and weighty metals from and anticipation of biofouling in water bodies, outstandingly in inland aquaculture (Iwuozor et al. 2021). The basic attributes of nanostructures that work with this utilization are their sizes, similarly high steadiness, and processing ability. Zinc oxide NPs, iron oxide NPs, tin oxide NPs, silver NPs, and carbon nanotubes (CNTs) are probably the most read-up gatherings of nanomaterials for use in aquaculture water quality administration (Ighalo et al. 2021). Additionally, their presentation has been accounted for to significantly improve in occurrences where they are utilized synergistically with biopolymers, for example, green growth, by taking advantage of their inborn surface hydrology and photosynthetic capacity (Hesni et al. 2021).

### 5.2.7 Nanomaterials in Reducing Heavy Metals

Iron oxide NPs are extremely appealing due to their small size, high surface-to-area proportion, surface modifiability, outstanding attractive characteristics, and biocompatibility, as demonstrated by Hesni's experiments with the nanoparticle and *Chlorella vulgaris* (Ighalo et al. 2021; Xu et al. 2012). Because of these qualities, iron oxide particles have proven to be quite effective in removing heavy metals from amphibian wastewater. For example, the most severe retention limit for Pb particles by Fe3O4 NPs in a concentrate was 360 mg/g (Nassar 2010), which is higher than recently disclosed characteristics for less expensive adsorbents. To attain the highest limit, a few factors in the batch adsorption cycle were adjusted, including pH, temperature, the grouping of initial Pb particles, and coincident particles. The iron NPs' small size was an important factor in the growth of metal particles from their placement onto metal sites on the adsorbents. $Fe_3O_4$ NPs were found to be extremely feasible and economical for disposing of and recovering metals in inland amphibian effluents in this way. In a similar vein, a few functionalized materials, such as chelating ligands, have been used to improve the surface modification of nanomaterials to improve their capability for heavy metal evacuation using carbon-encapsulated attractive nanoparticles (CEM NPs), which achieved 95 percent take-up for both ions. The adsorption isotherms for the metals produced adsorption limits, a significant property that determines how much an adsorbent is required for the removal of a specific measure of toxin, that is somewhere between 1.23 and 3.21 mg/g for the two metals, which is significantly higher for initiated carbon. Essentially, $NH_2$-adjusted $Fe_3O_4$ NPs were developed for the ejection of $Cu^{2+}$ particles that occurred between the nanoparticles and the amine lot on the altered nanomaterial's outer layer (Bystrzejewski et al. 2009).

This confirms the possibility that modified NPs may perform better in water quality management applications than unmodified NPs. Heavy metals in amphibian biota can also be removed using natural-source NPs, which have the advantage of being more effective due to their biodegradability. After estimating using a Nuclear Assimilation Spectrophotometer, the performance of Vetiveriazizanioides nanoadsorbents in reducing the heavy metal substance revealed that the reduction of the metals is in the range of decreases like Hg > As > Disc > Pb (Kumar et al. 2021).

## References

Aklakur M, Rather MA, Kumar N (2016) Nanodelivery: an emerging avenue for nutraceuticals and drug delivery. J Crit Rev Food Sci Nutr 56:2352–2361

Bystrzejewski M, Pyrzyńska K, Huczko A, Lange H (2009) Carbon-encapsulated magnetic nanoparticles as separable and mobile sorbents of heavy metal ions from aqueous solutions. Carbon 47(4):1201–1204

Deilamy Pour H, Mousavi SM, Zakeri M, Keyvanshokooh S, Kochanian P (2021) Synergistic effects of selenium and magnesium nanoparticles on growth, digestive enzymes, some serum biochemical parameters and immunity of asian sea bass (*Lates calcarifer*). Biol Trace Elem Res 199:3102–3111

Diallo, M.; Brinker, C.J. Nanotechnology for sustainability: environment, water, food, minerals, and climate. In Nanotechnology research directions for societal needs in 2020: retrospective and outlook; Roco, M.C., Hersam, M.C., Mirkin, C.A., eds.; Springer: Dordrecht, 2011; pp. 221–259

Fathi M, Mozafari MR, Mohebbi M (2012) Nanoencapsulation of food ingredients using lipid based delivery systems. Trends Food Sci Technol 23:13–27

Hesni MA, Hedayati A, Qadermarzi A, Pouladi M, Zangiabadi S, Naqshbandi N (2021) Using Chlorella vulgaris and iron oxide nanoparticles in a designed bioreactor for aquaculture effluents purification. Aquac Eng 90:102069

Ighalo JO, Sagboye PA, Umenweke G, Ajala OJ, Omoarukhe FO, Adeyanju CA, Ogunniyi S, Adeniyi AG (2021) CuO nanoparticles (CuO NPs) for water treatment: a review of recent advances. Environ Nanotechnol Monit Manage 2021(15):100443

Iwuozor KO, Ighalo JO, Ogunfowora LA, Adeniyi AG, Igwegbe CA (2021) An empirical literature analysis of adsorbent performance for methylene blue uptake from aqueous media. J Environ Chem Eng 9(4):105658

Joseph L, Jun BM, Flora JRV, Park CM, Yoon Y (2019) Removal of heavy metals from water sources in the developing world using low-cost materials: a review. Chemosphere 229:142–159

Kailasam M, Thirunavukkarasu AR, Chandra PK, Pereira S, Rajendran KV (2006) Induction of maturity and spontaneous spawning of captive brood stock of bhetki, *Lates calcarifer* (Bloch) through hormonal manipulation. In: Singh BN, Pandey AK (eds) Recent advances in hormonal physiology of fish and shell fish reproduction. Ms. Narendra Publishing House, New Delhi, pp 185–195

Karamzadeh M, Yahyavi M, Salarzadeh A, Nokhbe Zare D (2021) The effects of different concentrations of selenium and zinc nanoparticles on growth performance, survival and chemical composition of whiteleg shrimp (*Litopenaeus vannamei*). Iran Sci Fish J 29:43–51

Kumar N, Singh NP (2019) Effect of dietary selenium on immuno-biochemical plasticity and resistance against aeromonas veronii biovar sobria in fish reared under multiple stressors. Fish Shellfish Immunol 84:38–47

Kumar VS, Vidhyasri R, Dhevasenaa P, Ganeshbabu O (2021) Reduction of toxic elements in marine environment using Nano adsorbent materials. Mater Today Proc 37:1554–1557

Kumari R, Gupta S, Singh AR, Ferosekhan S, Kothari DC, Pal AK, Jadhao SB (2013) Chitosan nanoencapsulated exogenous trypsin biomimics zymogen-like enzyme in fish gastrointestinal tract. PLoS One 8(9):e74743. https://doi.org/10.1371/journal.pone.0074743

Mahdave Jehanabad J, Rastiannasab A, Ghaedi A, Mahmodi R, Salahi Ardakani MM (2019) Effect of different levels of selenium nanoparticles on some reproductive indices in rainbow trout (*Oncorhynchus mykiss*). Aquat Physiol Biotechnol 7:113–132

Meliana Y, Harmami SB, Restu WK (2017) Characterization of nanoencapsulated *Centella asiatica* and *Zingiber officinale* extract using combination of malto Dextrin and Gum Arabic as matrix. IOP Conf Ser Mat Sci Eng 172(012065):1–7

Nassar NN (2010) Rapid removal and recovery of Pb (II) from wastewater by magnetic Nano adsorbents. J Hazard Mater 184(1–3):538–546

Nuttall KL (2006) Evaluating selenium poisoning. Ann Clin Lab Sci 36:409–420

Sarkar B, Bhattacharjee S, Daware A, Tribedi P, Krishnani KK, Minhas PS (2015) Selenium nanoparticles for stress-resilient fish and livestock. Nanoscale Res Lett 10:371

Sethi SN, Panigrahi A (2011) Innovative aquaculture. Fish Chimes 31(3):57–58

Shah BR, Mraz J (2020) Advances in nanotechnology for sustainable aquaculture and fisheries. Rev Aquac 12:925–942

Toranzo AE, Magarinos TB, Romalde JL (2005) A review of the main bacterial fish diseases in mariculture systems. Aquaculture 246:37–61

Verma D, Gulati N, Kaul S, Mukherjee S, Nagaich U (2018) Protein based nanostructures for drug delivery. J Pharm 2018:1–18

Xu P, Zeng GM, Huang DL, Feng CL, Hu S, Zhao MH, Lai C, Wei Z, Huang C, Xie GX, Liu ZF (2012) Use of iron oxide nanomaterials in wastewater treatment: a review. Sci Total Environ 424:1–10

Zahid M, Rashid A, Akram S, Rehan ZA, Razzaq W (2018) A comprehensive review on polymeric nano-composite membranes for water treatment. J Membr Sci Technol 8:1–20

Zodrow K, Brunet L, Mahendra S, Li D, Zhang A, Li QL, Alvarez PJJ (2009) Polysulfone ultrafiltration membranes impregnated with silver nanoparticles show improved biofouling resistance and virus removal. Water Res 43:715–723

# Chapter 6
# Introduction of Nanotechnology Intervention in Aquaculture

**A. Sai Ramesh, Ayurika Tripathy, N. V. P. Meghana, L. Karthik, G. Kiran, and Vishnu Kirthi Arivarasan**

## 6.1 Introduction to Nanotechnology

In the word nanotechnology, the prefix "nano" signifies a billionth, and any innovation working at the nanoscale and having applications in reality can be named as nano innovation. However, there is no ideal meaning of nanotechnology; it tends to be characterized in numerous potential ways, or in basic words, it is simply "innovation occurring at the nanoscale" (Kaehler 1994), where the nanoscale is essentially the range between 1 and 100 nm. Nanoscience and innovation have detonated in prominence in the earlier decade, attributable to the accessibility of novel advancements for assembling nanomaterials, as well as different factors, for example, portrayal and control apparatuses (Daniel and Astruc 2004; Rao and Cheetham 2001).

A. S. Ramesh (✉) · A. Tripathy · N. V. P. Meghana
Department of Biotechnology, Vel Tech High Tech Dr. Rangarajan Dr. Sakunthala Engineering College, Chennai, Tamil Nadu, India

L. Karthik
Synthetic Biology, Research & Department, Salem Microbes Pvt Ltd, Salem, Tamil Nadu, India

G. Kiran
School of Pharmacy, Anurag University, Hyderabad, Telangana, India

V. K. Arivarasan (✉)
Department of Microbiology, Lovely Professional University, Phagwara, Punjab, India



V. K. Arivarasan et al. (eds.), *Nanotechnological Approaches to the Advancement of Innovations in Aquaculture*, Nanotechnology in the Life Sciences, https://doi.org/10.1007/978-3-031-15519-2_6

## 6.2 Importance of Nanoscale Materials

There are many reasons why nanotechnology has become a very important field and has created many new opportunities, and the factors of nanoscale materials responsible are listed below (National Science and Technology Council 2000):

### *6.2.1 The Quantum Size Effect*

Numerous nanoscale components have an upgraded proportion of surface region to volume, because of which new quantum mechanical impacts are given. This impact can modify the actual properties, for example, the liquefying point, changing the limit without changing the compound arrangement of materials (National Science and Technology Council 2000; Abad 2005).

### *6.2.2 Catalysis*

Due to the large surface area and volume ratios, chemical properties at the nanoscale provide an advantage for processes like catalysis, energy storage, drug delivery systems, etc. (Mansoori and Soelaiman 2005).

### *6.2.3 Structural Organization*

Nanotechnology has allowed us to put man-made nanoscale objects into living cells (Hu et al. 1999). It has provided the possibility to explore the microstructure and macrostructure of matter using molecular self-assembly. This certainly is a very powerful instrument in materials science (Whitesides 2005).

### *6.2.4 Enhanced Mechanical Properties*

Macroscopic systems made up of nanostructures have a number of improved mechanical properties, such as the improved strength and hardness of nanomaterials and lightweight nanocomposites. Such systems can potentially have their molecular structures altered nanomechanically (Bhushan 2010).

## 6.3 History of Nanotechnology

People have been examining "nano"-measured presence for quite a while. Dr. Richard P. Feynman (Ghorbanpour et al. 2015), a physicist and Nobel Laureate, is broadly credited for begetting the expression "nanotechnology." Each of his logical talks is viewed as exemplary. "There's A lot of Room at the Last: A Challenge to Enter New Field of Physical science" was the title of one of his well-known displays. As Feynman referenced in his talks, the principles of nature and our capacity to work at the nuclear and subatomic levels didn't restrict us, yet it was our absence of satisfactory methodologies and gear that held us back from accomplishing our objectives. He considered "scaling down by vanishing," which he characterized as meager film arrangements, "making small things with molecules," like smaller than usual machines, electronic circuits, and industrial facilities; and "tackling the issues of iotas," which he characterized as "fixing the issues of particles. Nanorobotics is being utilized to provide "grease." He additionally exhorted "revamping molecules" to tackle the issue and make a wide scope of nanostructures and nanodevices. Applications that have come about because of Feynman's discourse in 1959 include the control of single molecules on a silicon surface, the catching of single, 3D, nanometer-width colloidal particles, the control of single, 3D, nanometer-measurement colloidal particles, electrostatic techniques, and the position of single particles. A checking burrowing magnifying instrument was utilized to analyze particles (STM) (Mansoori et al. 2005).

## 6.4 Methods of Nanofabrication in Nanotechnology

There are two nanotechnological approaches, i.e., the top-down approach and atomic-scale fabrication representing the bottom-up approach (Mansoori et al. 2007; Ghorbanpour and Hatami 2015), which involves the manufacture of device structures through the systematic self-assembly of molecules, atoms, or other basic units of matter.

### 6.4.1 *Top-Down Approach*

Top-down manufacturing is the process by which bulk materials are broken down into smaller and smaller pieces via mechanical, chemical, or other forms of energy. The electronics industry is the most successful at manipulating these fragments to create functional structures at the micro- and nanoscale. This approach has led engineers and physicists to manipulate progressively smaller fragments of matter via photolithography and other techniques (Lee et al. 2010).

### 6.4.2 *Bottom-Up Approach*

The bottom-up approach represents the self-assembly of atoms, molecules, and machines from basic building blocks to the making of nanomaterials and nanodevices. Such an approach of making nanostructures atom by atom using carefully controlled chemical reactions will lead to the generation of cheaper techniques compared with lithographic methods (Lee et al. 2010).

In order to develop objects that are actually artificially and organically steady, nanotechnology can be described as a nuclear or subatomic method. It requires the estimation, forecasting, and manufacture of issues based on the size of ions and atoms. In the event that nanotechnology keeps on developing at its present rate, it will affect the existence of almost everybody in the world in the next few years (Lee et al. 2010).

## 6.5 Introduction to Aquaculture

The status and possibilities of aquaculture are found with regard to a bigger food climate that incorporates wild amphibians and earthbound food sources. The thinking and asset base needed for aquaculture advancement are analyzed considering cultural advancement, social inclinations, and human necessities. The lopsided turn of events and current pertinence of aquaculture all over the planet, as well as its huge fluctuation of structure and capacity when contrasted with earthly creature creation, are featured. The new drivers of interest and creation development are explored, as well as the tireless connections between wild stock abuse, full life cycle culture, and various transitional structures (FAO 2018).

Aquaculture, ordinarily known as aqua farming, is the quickest-developing food area on earth. This increment is due to rising protein interest and a consistently expanding human population. The five essential aquaculture markets are anticipated to reach $87.6 billion by 2025, as indicated by partnered statistical surveying, with a build-year development pace of 4.9 percent from 2018 to 2025. The USA, Europe, China, Russia, and Japan are the five biggest aquaculture markets. In spite of the way that aquaculture has been around for a very long time, it is as yet a youthful and extending area. It can provide some significant experience from animal cultivation and faces various issues such as infection prevention, low-sway creation, feeds, and nourishment. Trendsetting innovation, then again, has helped out. The Internet of Things (IoT) has productive potential for further developing effectiveness and keeping up with seagoing species' well-being (Mirsasaani et al. 2013).

The repercussions for the area with quick-extending feed needs are analyzed, as is an arising pattern for getting hydroponics taken care of from options in contrast to marine materials. Hydroponics is turning into a significant stock of marine parts, and nontraditional and inventive feed fixings, which are normally imparted to earthbound creatures, are examined. The ramifications for hydroponics' anticipated supported development are examined with regard to reasonable growth, with the issues that regular heightening and eminent mix inside and between esteem chains face being researched. The appraisal proceeds with a discussion of the ramifications for subordinate jobs as well as assessments for potential fates dependent on limited assets and expanding demand (Ghorbanpour et al. 2017).

The rearing, developing, and collecting of fish, shellfish, green growth, and different living beings in a wide range of water living spaces is known as hydroponics.

Innovation has made it conceivable to develop food in beachfront marine waters and the vast sea as the interest in fish has developed. Aquaculture is a way for creating food and other business things, as well as reestablishing living space, renewing wild stocks, and revamping undermined and imperiled creature populaces. Aquaculture is divided into two kinds: marine and freshwater. NOAA's exercises are generally centered around marine hydroponics, which alludes to the cultivation of marine and estuarine creatures.

Freshwater hydroponics in the USA produces species like catfish and trout. Freshwater hydroponics is, for the most part, done in lakes or other counterfeit frameworks.

NOAA is devoted to advancing a monetarily, environmentally, and socially practical hydroponics business. NOAA researchers and colleagues are attempting to all the more likely comprehend the natural effects of hydroponics in different settings and to foster ideal administration practices to assist with restricting the chance of unfortunate results.

### 6.5.1 *The Upsides of Aquaculture*

In the course of the last decade, hydroponics, or fish cultivation, has acquired ubiquity in the United States as a feasible method for creating fish. Expanding interest in new fish, as indicated by certain specialists, has come down on normal populaces. Hydroponics, or the cultivation of marine fish and shellfish, is turning out to be progressively famous for providing this interest. Hydroponics is vital for the economy since it utilizes a large number of individuals for tasks and assistant administration.

Worldwide fisheries send out as of now more money than any other exchanged agrarian item in the world, including rice, cocoa, and espresso, as indicated by the Natural Safeguard Asset, a non-benefit ecological association (http://agrilinks.org/post/advanced-technology-and-aquaculture-drones-and-artificial-intelligence-make-difference) (Fig. 6.1).

**Fig. 6.1** Schematic representation of intervention by nanotechnology and aquaculture

## 6.6 Intervention of Nanotechnology and Aquaculture

Hydroponics has emerged as one of the world's quickest-developing food enterprises lately, assisting with fulfilling the developing need for animal protein. Notwithstanding, illness predominance, synthetic contamination, and ecological factors all play a role in the elements that essentially affect helpless feed usage and mental degeneration and break the area's capacity to contribute to worldwide food security. In this situation, to manage these issues, new courses in science and innovation have been cleared. To restrict these downsides in this hydroponics area, "nanotechnology" has emerged as a distinct advantage. With inventive nanotools, there is colossal potential to further develop hydroponics. This investigation is critical (Ghorbanpour et al. 2017). There is a critical hole in specialized advancement for drug use, infection treatment, water quality control, and item improvement. Considering an illustration of a specially custom-made fish (Shah and Mraz 2020) for further developed well-being and creation, better well-being is energized by epigenetic and nutrigenomic cooperations. Rearing accomplishment through productive development conveyance and nutraceutical organization for fast bringing forth, generating invigorating specialist powerful development advancement and culture time decrease. Around here, auto-transgenics and powerful antibodies are being utilized. To beat these impediments, grasping, coordinating, and consolidating approaches as well as carrying out new logical and innovation are required. It is basic to support a positive hydroponics. Right now, the hydroponics business is going through new logical and innovative advancements to create more talented laborers.

The ceaseless and striking development in the stock of ocean food varieties for human utilization is reflected in the increment in worldwide fish creation, with around 171 million tons in the year 2016. The creation flood is worth USD 362 billion, accounting for USD 232 billion from hydroponics creation, and out of which around 47% of the all-out addresses hydroponics and almost 53% utilizes nonfood (counting decrease to fishmeal and fish oil). The expansion underway isn't viewed as the sole component of increased utilization of ocean food varieties; however, certain variables like lessening wastage through appropriate bundling material are worth considering. Besides, the commitment of hydroponics and ocean food varieties towards sanitation and security can be a promising support point concerning the 2030 agenda (Ferosekhan et al. 2014). There are many difficulties in accomplishing the maintainability and set point of the 2030 plan, and one related angle includes the expansion in overall fish creation to around 171 million tons in 2016 mirrors the ceaseless and marvelous development in the inventory of marine food varieties for human use. The increment underway is about USD 362 billion. Hydroponics creation produces USD 232 billion, and hydroponics represents generally 47% of the aggregate and generally 53% of nonfood applications (remembering a decrease for contamination) (including fishmeal, as well as fish oil). The ascent in yield is anything but decent. It is viewed as the sole reason for higher utilization. Notwithstanding certain conditions, for example, a decrease in wastage through the utilization of proper bundling materials is something to contemplate.

Moreover, a job in aquaculture and the ocean food varieties that add to sanitation and security can be a promising choice. Part with connection to the 2030 plan. There are various hindrances to overcome to accomplish this drawn-out objective of the 2030 Plan for a Reasonable Turn of Events, and one related part involves. On account of causes connected with decaying, the nature of fish is turning into a major pressing issue in the fish industry all over the planet. Notwithstanding mechanical improvements in marine food creation, fish and fish products have become progressively inclined to dismissal because of the globalization of food exchange because of their uncommonly short-lived nature and low unrefined substance quality (Ferosekhan et al. 2014).

One of the innovative leaps forward was the bundling of detached fish items to shield them from microbial weakening, parching, and oxygen. Alternate clear bundling was used first, followed by changed air bundling with regard to the previously mentioned bundle purposes (Guide). Because of the hardships of poison development by *Clostridium botulinum* and temperature misuse, the United States has extremely strict laws encompassing the utilization of Guide. The time-temperature markers (TTIs) utilized in Guide, then again, are the most probable solution for these problems (Aklakur et al. 2016).

### 6.6.1 Current Nanotechnological Methods in Aquaculture

Aquaculture, being the fastest-expanding food sector, has the potential to help people retain their socioeconomic status. It has the potential to make a big contribution to society's food and nutrition by delivering highly useful aquatic proteins and fats. Aquaculture is one of the fastest-growing food processing industries, owing to the growing demand for seafood and fish around the world, as well as the fact that continued growth of aquaculture is seen as a vital strategy for ensuring universal food and nutritional safety. Because of their composite composition and active environment, where each element varies quickly and determinedly, various food products are particularly ephemeral.

### 6.6.2 Nanotechnology as a Part of Aquaculture and Fisheries

Nanotechnology can transform the fisheries and aquaculture industries by providing new tools such as rapid disease diagnosis and improving fish's ability to absorb medications such as hormones, vaccinations, and nutrients quickly. According to the National Science Foundation (USA), the worldwide nanotechnology sector is expected to reach a value of USD 1 billion by 2025. By 2015, one trillion dollars will have been spent. This may be achievable because of nanotechnology's huge potential in fields other than electronics and materials research, like human health, animal food, and agriculture sectors. For instance, aquaculture is employed in cancer treatment, the development of nonviral vectors for gene therapy, biomedical and biological research, technologies for biomolecular analysis, and drug delivery (Lampila and McMillin 2012).

Targeting medication delivery, clinical diagnosis, and treatments, among other things, use DNA, proteins, or cells. Despite the fact that much of the development research is conducted in laboratories, there are various glimpses of the future needed to boost the possible application of nanotechnology in aquaculture (http://aquafind.com/articles/Nanotechnology-In-Aquaculture.php).

### 6.6.3 Nanoparticles for Enhancement and Fish Growth

When immature carp and sturgeon were fed iron nanoparticles, their growth rates increased by 30% and 24%, respectively, according to researchers from the Russian Academy of Sciences. Different selenium sources (nano-Se and selenomethionine) supplied in the basal diet improved the final weight, relative gain rate, antioxidant status, Glutathione Peroxidase (GSH-Px) activities, and muscle Se concentration of crucian carp, according to research (Carassius auratus gibelio). Furthermore, nano-Se was found to be more effective in raising muscle selenium levels than organic selenomethionine. Similarly, the nanolevel administration of these nutraceuticals improved the growth and performance of the investigated fishes (https://world-oceanreview.com/en/wor-2/aquaculture/).

### 6.6.4 Vaccine Delivery

Vaccines have played an important role in aquaculture as a defense mechanism against viruses, protecting the host from diseases caused by these pathogens. Oral vaccination or injectable vaccination are the most dependable and successful methods of vaccination in fisheries. The latter, a typical adjuvant approach, necessitates the preparation of vaccines using oil/water formulations, which have numerous negative consequences (Rather et al. 2011).

### 6.6.5 Nutrient Distribution

Nutraceuticals are notable for their capacity to assist fish with increasing their turn of events and immunological attributes. In any case, rather than addressing fundamental requirements, their execution requires bigger costs. Accordingly, outrageous alerts ought to be practiced in their utilization to diminish waste and upgrade productivity (Adversary (Companions of the Earth) 2008) (Ghorbanpour et al. 2017). There is an enormous group of proof supporting nanotechnology's capacity in the powerful organization of dietary enhancements and nutraceuticals in fisheries. These frameworks are intended to work on the bioavailability, and, in this way, the viability of supplements by expanding their dissolvability and shielding them from the unfriendly stomach climate. As a general rule, these kinds of arrangements, along with the organizational procedures, can bring about fish mortality. To resolve these issues, researchers have fostered a nano conveyance framework as a substitute method for immunization conveyance in fish, which is believed to be more secure as well as more powerful. Until now, different epitome approaches have been created and tried in this specific situation. Among these, alginate particles were distinguished as possible contenders for oral immunization organization in oceanic species (Zhou et al. 2009). Alginate is a copolymer of b-D-mannuronic corrosive (M) and a-L-guluronic corrosive (G) found in an assortment of earthy-colored green growth species and as a polysaccharide in microbes. It has a notable mechanical ability (Abdul Kader Mydeen and Haniffa 2011). The utilization of nutraceuticals in fish and shellfish for well-being across the board, esteem expansion, and stress relief is another area of hydroponics study. Regardless of their unassuming need, nutraceuticals have a more noteworthy expense of consolidation.

Subsequently, it should be utilized so that waste is limited to augment effectiveness and make the final result monetarily doable. The advancement of a nanodelivery strategy for these mixtures could help address a portion of the issues with their business usage in hydroponics. The utilization of nanoparticles to convey nutraceuticals in fish feed and neutrogenomics studies has huge potential. Moreover, various feed nanoformulations help maintain feed consistency and taste (Rather et al. 2011).

### *6.6.6 Water Purification*

One of the most critical pillars for long-term aquaculture is water treatment. In recent years, water contamination has been identified as the world's most serious health threat, and it is steadily increasing as a result of waste dumping from cities, businesses, and agriculture, as well as the misuse of antibiotics and other synthetic substances in fisheries. The degradation of waters in this way has a direct impact on human health by reducing the availability of clean groundwater, but it also has an indirect impact on aquatic creatures, whose ingestion can result in a variety of foodborne illnesses. Apart from that, the fishery industry suffers a significant financial loss as a result of microorganisms and heavy metals in these waters, which induce growth retardation and fish death. Nanotechnology has important applications in aquaculture, such as water treatment to offer a favorable and safe environment for fish spawning. In this light, the scientific community supports adsorption and photocatalysis as the most efficient and cost-effective methods of water purification (Ghorbanpour et al. 2017).

Apart from these applications of nanotechnology, nanotechnology-based materials and products are known to have negative effects on the environment and human health, in addition to their fundamental application in the development and sustainability of aquaculture. Aquatic creatures, in particular, are at a higher risk of exposure to the potential toxicity of these compounds due to their greater sensitivity. As a result, it is critical to take into account the negative and harmful impacts of nanomaterials on aquatic creatures (Joosten et al. 1997) (Fig. 6.2).

Altair Nanotechnologies, established in Nevada, develops Nano-Check, a water-cleaning solution for swimming pools and fishponds. It employs 40 nm lanthanum-based particles that absorb phosphates from the water and prevent algae growth. Nano-Check is currently being tested in swimming pools on a wide basis, and Altair released a swimming pool cleaner in early 2005. Altair hopes to see Nano-Check used in tens of thousands of commercial fish farms throughout the world, where algae and heavy metal removal and avoidance are now prohibitively expensive. Altair plans to expand its experiments to confirm its effect on fish, as well as the effects of nanoparticle-laden runoff on human health and the environment, according to the business. Furthermore, nanoscale weedicide and soil-wetting agent distribution could be particularly effective for aquatic weed management in large water bodies, as well as stress mitigation owing to climate change and aquatic pollution (http://aquafind.com/articles/Nanotechnology-In-Aquaculture.php).

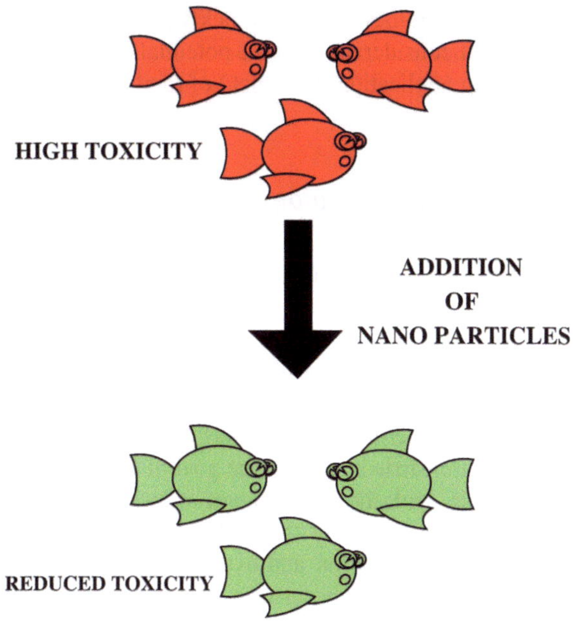


**Fig. 6.2** Differential description of toxicity by nanoparticles Nanotechnology Devices for Aquatic Environment Management

### *6.6.7 Nanotechnological Interventions in Food Processing*

The food business has effectively embraced an emerging approach like food nanotechnology at the turn of the twenty-first century. Nanotechnology is the study of controlling matter at the nuclear and subatomic levels with no less than one aspect estimated in nanometers. It is an expansive innovation, similar to biotechnology and data innovation, that is fused into a bigger more specialized framework. Besides, it can possibly change agribusiness and other related fields, like hydroponics and fisheries. It very well might be utilized to make new and progressed apparatuses and systems in an assortment of fields, including hydroponics, fish biotechnology, fish regeneration, fish hereditary qualities, and amphibian well-being, to give some examples. It's only recently that they've been utilized in the hydroponics and food-handling industries. Albeit fruitful nanotechnology interventions in this field are as of now restricted, numerous food researchers have perceived the colossal capability of nanotechnology to lead all food handling businesses (Gombotz 1998).

Food fabrication organizations should foster new innovations to be serious in the food handling area and proposition new, proper, tasty, and safe food items. One of the most fundamental objectives of food handling is to broaden the timeframe of realistic usability of the item while additionally maintaining its newness and quality.

Nanotechnology is a significant spearheading innovation in the contemporary period, with a wide scope of uses in the food handling industry [43].

Nanotechnology has been quickly progressing in current times, and it has drawn in a ton of consideration for the potential outcomes of nanoscale bundling. It has an animating application in the field of food and related things.

### 6.6.8 Nanotechnology in Sea Food Preservation

Fish preservation is a crucial part of the fisheries' processing process. Typically, fish farms or other fish catching locations are placed far from retail outlets, resulting in a high rate of fish decomposition and the uncertainty of their sale in the market. When a large quantity of fish is harvested in excess of market demand, it is necessary to preserve and process the fish for future use. As a result, preservation and processing are critical components of commercial fisheries.

They are carried out with the goal of keeping the fish fresh and safe for as long as possible, with the least amount of loss in vital quality features, nutritional content, and digestibility of the flesh. Many preservation procedures are necessary to avoid fish deterioration and extend the shelf life of seafood. These methods are intended to prevent the growth and proliferation of spoilage microorganisms, as well as metabolic changes that lead to a loss of fish quality.

Furthermore, the safety of seafood is jeopardized as the human population grows. These difficulties highlight the need to preserve seafood quality. Despite the fact that traditional preservation techniques such as irradiation prevent the growth of microbes in fish and despite the relevance of current conservation methods throughout the storage period, patrons in recent years have favored the use of various food preservation methods as these methods would have unfavorable effects on human health (Li et al. 2008; Dar et al. 2020).

Experts have been on the lookout for novel alternatives as a result of the fears. As a result, an enormous effort is currently being made to develop new antimicrobial compounds with the goal of regulating microbial infection in seafood. In recent years, replacement strategies for removing food stability have been anticipated. In this regard, nanotechnology in regard to food processors' concerns has proven to be an effective innovative tool for increasing food stability and thus being functionalized in order to keep color and also boost flavor (Ikape 2017). As previously stated, seafood yield is particularly delicate, so the use of nanotechnology to address problems in dealing with related concerns is extremely important. Human health has been shown to be at risk from histamine, halophytic pathogenic bacteria, and parasites in traditionally developed fish and other marine food products. Flavorings on a nanoscale could be used to change a product's textural qualities, shelf life, and nutritional profile, as well as to differentiate food pathogens and serve as food standards indicators. Food packaging nanotechnologies are mostly used to extend product shelf life, identify spoiled components, or heighten product prominence by restricting gas flow over the product's packaging (Tassou et al. 1996; Rather et al. 2011).

Fish preservation is a very important processing aspect of fisheries. Typically, the fish farms or other fish-capturing sites are located at great distances from retail points, and there is a high frequency of fish decomposition and the uncertainty of their sale in the market. When the fish is harvested in large quantities, greater than the market requirement, their preservation and processing become necessities for their future application. Preservation and processing therefore are indispensable components of commercial fisheries.

They are performed with the purpose of ensuring that the fishes remain fresh and safe for a long time with a minimum loss in essential quality attributes, nutritive value, and the digestibility of their flesh. To prevent fish spoilage and lengthen the shelf life of seafood, many preservation techniques are required. These techniques are designed to inhibit the growth and proliferation of spoilage bacteria and the metabolic changes which result in the loss of fish quality (Perreault et al. 2015).

Fish preservation is a crucial part of the fisheries processing process. Typically, fish farms or other fish-catching locations are placed far from retail outlets, resulting in a high rate of fish decomposition and the uncertainty of their sale in the market. When a vast quantity of fish is captured, much beyond the market need, preservation and processing become essential for their future use. As a result, preservation and processing are critical components of commercial fisheries. They are carried out with the goal of keeping the fish fresh and safe for as long as possible, with minimal loss of vital qualitative features, nutritional content, and digestibility of the flesh. Many preservation procedures are necessary to avoid fish deterioration and extend the shelf life of seafood. These methods are intended to prevent the growth and proliferation of spoilage microorganisms, as well as metabolic changes that lead to a reduction in fish quality. As previously stated, seafood yield is particularly delicate, so the use of nanotechnology to address problems in dealing with related concerns is extremely important.

On the other hand, literature reviews on the use of nanotechnology for the preservation of seafood reveal that it is restricted to the use of nanoparticles and nanoemulsions. The use of nanofibers as a replacement for preserving fish products is possible. Based on the scientific fact that nanofibers can be used to encapsulate a variety of antimicrobial composites in order to prevent microbial infectivity and disease, nanofibers may be capable of reducing microbial activity. In reverence, a novel strategy to block the formation of microorganisms in fish meat would be to electro-spin nanofibers with antibacterial capabilities on the external surface layer of the fish (Kushnerova et al. 2010; Chellaram et al. 2014).

Because of their differentiating properties, environmental compatibility, and chemical individuality, biosynthetic nanomaterials have gotten a lot of interest in the current period in a range of sectors of natural science.

### *6.6.9 Nanoemulsions*

Nanoemulsions have remarkable stability and physicochemical qualities due to their droplet size of less than 100 nm. Because of their smaller size, lipophilic substances have a greater surface area per unit of mass, resulting in increased biological activity. Nanoemulsions open up new possibilities for creating next-generation edible films (Ozogul et al. 2017; McClements and Rao 2011; Solans et al. 2005; Sekhon 2014).

### *6.6.10 Nanosensors*

Nanosensors are currently being used in nanotechnological hardware for cow assessment and fishpond cleaning. Fish lakes are given DNA nanovaccines as nanocapsules containing short strands of DNA, which are caught in the fishes' cells. The captured containers are then broken with ultrasounds, permitting the DNA to be delivered and the fish to foster an immunological reaction. This advancement innovation is rapidly becoming legitimate in the utilization of iron nanoparticles to speed up fish development and a customized drug discharge framework.

In the hydroponics business, nanotechnological biosensors can likewise be utilized for microbial control. The Public Flight and Space Organization has made a carbon nanotube-based delicate biosensor fit for distinguishing minute measures of microorganisms like microbes, infections, and parasites, as well as weighty metals in water and food varieties. The nanocolloidal bit is perhaps the main nanotechnology item, acting as a catalyst on a wide scope of microscopic organisms, growths, parasites, and infections by changing over a compound for use in their metabolism (Otoni et al. 2016).

Nanotechnology is anticipated to provide mechanical intercessions in hydroponics and marine food creation, handling, stockpiling, transporting, recognizing, well-being, and security. In any case, before being popularized, nanotechnology-determined items should exhibit their financial possibility. Data on the financial seriousness of nanotechnology-determined items has been scarce up to now.

Nanotechnology can possibly be a distinct advantage in food handling and safeguarding. Infection episodes in fish could be effectively controlled with the consideration of nanotechnological instruments, for example, nanocomposites, nano biosensors, nano muds, and nanovaccines (Gombotz 1998).

## References

Abad E (2005) Nano dictionary. Collegium Basilca, Basel

Abdul Kader Mydeen K, Haniffa MA (2011) Evaluation of antibacterial activity of medicinal plants on fish pathogen Aeromonas hydrophila. J Res Biol 1:1–5. Abebe LS, Su Y-H, Guerrant RL, Swami NS, Smith JA (2015)

Aklakur M, Asharf Rather M, Kumar N (2016) Nanodelivery: an emerging avenue for nutraceuticals and drug delivery. Crit Rev Food Sci Nutr 56:2352–2361

Bhushan B (ed) (2010) Springer handbook of nanotechnology. Technology & Engineering, Springer Science & Business Media

Chellaram C, Murugaboopathi G, John AA, Sivakumar R, Ganesan S, Krithika S, Priya G (2014) Significance of nanotechnology in food industry. APCBEE Procedia 8:109–113. https://doi.org/10.1016/j.apcbee.2014.03.010

Daniel MC, Astruc D (2004) Gold nanoparticles: assembly, supramolecular chemistry, quantum size-related properties, and applications toward biology, catalysis, and nanotechnology. Chem Rev 104(1):293–346

Dar AH, Rashid N, Majid I, Hussain S, Dar MA (2020) Nanotechnology interventions in aquaculture and seafood preservation. Crit Rev Food Sci Nutr 60(11):1912–1921. https://doi.org/10.1080/10408398.2019.1617232

FAO (2018) The state of world fisheries and aquaculture 2018- meeting the sustainable development goals. Rome. Licence. In: CC BY-NC-SA 3.0 IGO

Ferosekhan S, Gupta S, Singh A, Rather M, Kumari R, Kothari D et al (2014) RNA-loaded chitosan nanoparticles for enhanced growth, immunostimulation and disease resistance in fish. Curr Nanosci 10:453–464

Ghorbanpour M, Hatami M (2015) Changes in growth, antioxidant defense system and major essential oils constituents of Pelargonium graveolens plant exposed to nano-scale silver and thidiazuron. Indian J Plant Physiol 20(2):116–123

Ghorbanpour M, Hatami M, Hatami M (2015) Activating antioxidant enzymes, hyoscyamine and scopolamine biosynthesis of Hyoscyamus niger L. plants with nano-sized titanium dioxide and bulk application. Acta Agric Slov 105:23–32

Ghorbanpour M, Khanuja M, Varma A (eds) (2017) Nanoscience and plant-soil systems springer. Technology & Engineering, Soil Biology

Gombotz WR (1998) Protein release from alginate matrices. Adv Drug Deliv Rev 31:267–285

http://agrilinks.org/post/advanced-technology-and-aquaculture-drones-and-artificial-intelligence-make-difference

http://aquafind.com/articles/Nanotechnology-In-Aquaculture.php

https://worldoceanreview.com/en/wor-2/aquaculture/

Hu J, Odom TW, Lieber CM (1999) Chemistry and physics in one dimension: synthesis and properties of nanowires and nanotubes. Acc Chem Res 32(5):435–445

Ikape SI (2017) Fish spoilage in the tropics: a review. Octa J Biosci 5(2):34–37

Joosten P, Tiemersma E, Threels A, Dhieux-Caumartin C, Rombout J (1997) Oral vaccination of fish against Vibrio anguillarum using alginate microparticles. Immunol Aspects Oral Vaccin Fish 56:471–485

Kaehler T (1994) Nanotechnology: basic concepts and definitions. Clin Chem 40(9):1797–1799

Kushnerova NF, Fomenko SE, Sprygin VG, Kushnerova TV, Khotimchenko YS, Kondrat'eva EV, Drugova LA (2010) An extract from the brown alga Laminaria japonica: A promising stress-protective preparation. Russ J Mar Biol 36(3):209–214. https://doi.org/10.1134/S1063074010030077

Lampila LE, McMillin KW (2012) Major microbial hazards associated with packaged seafood. In: Advances in meat, poultry and seafood packaging. Woodhead Publishing, Sawston, pp 59–85

Lee J, Mahendra S, Alvarez PJJ (2010) Nanomaterials in the construction industry: a review of their applications and environmental health and safety considerations. ACS Nano 4(7):3580–3590

Li H, Zhang J, Wang T, Luo W, Zhou Q, Jiang G (2008) Elemental selenium particles at nano-size (Nano-Se) are more toxicto Medaka (Oryzias latipes) as a consequence of hyper accumulation of selenium: a comparison with sodium selenite. Aquat Toxicol 89:251–256

Mansoori GA, Soelaiman TAF (2005) Nanotechnology-an introduction for the standards community. J ASTM Int 2(6):17–38

Mansoori GA, Vakili-Nezhaad GR, Ashrafi AR (2005) Some mathematical concepts applicable in nanothermodynamics. Int J Pure Appl Math Sci 2:58–61

Mansoori GA, Mohazzabi P, McCormack P, Jabbari S (2007) Nanotechnology in cancer prevention, detection and treatment: bright future lies ahead. World Rev Sci Technol Sustain Dev 4(2–3):226–257

McClements DJ, Rao J (2011) Food-grade nanoemulsions: formulation, fabrication, properties, performance, biological fate, and potential toxicity. Crit Rev Food Sci Nutr 51(4):285–330. https://doi.org/10.1080/10408398.2011.559558

Mirsasaani SS, Hematia M, Tavasolid T, Dehkorda ES, Yazdia GT, Poshtiri DA (2013) Nanotechnology and nanobiomaterials in dentistry. In: Nanobiomaterials in clinical dentistry. Elsevier Inc.

National Science and Technology Council (2000) National nanotechnology initiative: leading to the next industrial revolution. National Science and Technology Council, Washington, DC
Otoni CG, Avena-Bustillos RJ, Olsen CW, Bilbao-Sainz C, McHugh TH (2016) Mechanical and water barrier properties of isolated soy protein composite edible films as affected by carvacrol and cinnamaldehyde micro and nanoemulsions. Food Hydrocoll 57:72–79. https://doi.org/10.1016/j.foodhyd.2016.01.012
Ozogul Y, Yuvka I, Ucar Y, Durmus M, Kosker AR, Öz M, Ozogul F (2017) Evaluation of effects of nanoemulsion based on herb essential oils (rosemary, laurel, thyme and sage) on sensory, chemical and microbiological quality of rainbow trout (Oncorhynchus mykiss) fillets during ice storage. LWT-Food Sci Technol 75:677–684. https://doi.org/10.1016/j.lwt.2016.10.009
Perreault F, De Faria AF, Nejati S, Elimelech M (2015) Antimicrobial properties of graphene oxide nanosheets: why size matters. ACS Nano 9(7):7226–7236. https://doi.org/10.1021/acsnano.5b02067. Nickols-Richardson SM, Piehowski KE (2008) Nanotechnology in nutritional sciences. Minerva Biotechnol 20(3):117
Rao CNR, Cheetham AK (2001) Science and technology of nanomaterials: current status and future prospects. J Mater Chem 11(12):2887–2894
Rather MA, Sharma R, Aklakur M, Ahmad S, Kumar N, Khan M, Ramya VL (2011) Nanotechnology: A novel tool for aquaculture and fisheries development. A prospective mini-review. Fish Aquac J 16(1–5):3. https://doi.org/10.4172/2150-3508.1000016
Sekhon BS (2014) Nanotechnology in Agri-food production: an overview. Nanotechnol Sci Appl 7:31. https://doi.org/10.2147/NSA.S39406
Shah BR, Mraz J (2020) Advances in nanotechnology for sustainable aquaculture and fisheries. Rev Aquacult 12(2):925–942. https://doi.org/10.1111/raq.12356
Solans C, Izquierdo P, Nolla J, Azemar N, Garcia- Celma MJ (2005) Nano-emulsions. Curr Opin Colloid Interface Sci 10(3–4):102–110. https://doi.org/10.1016/j.cocis.2005.06.004
Tassou CC, Drosinos EH, Nychas GJE (1996) Inhibition of resident microbial flora and pathogen inocula on cold fresh fish fillets in olive oil, oregano, and lemon juice under modified atmosphere or air. J Food Prot 59(1):31–34. https://doi.org/10.4315/0362-028X-59.1.31
Whitesides GM (2005) Nanoscience, nanotechnology, and chemistry. Small 1(2):172–179
Zhou X, Wang Y, Gu Q, Li W (2009) Effects of different dietary selenium sources (selenium nanoparticle and selenomethionine) on growth performance, muscle composition and glutathione peroxidase enzyme activity of crucian carp (Carassius auratus gibelio). Aquaculture 291:78–81

# Chapter 7
# Alarming Viral Pathogens in Shrimp Industry and Nanotechnology

**Hethesh Chellapandian, Mohamed Sheik Aadham, Mohamed Jaabir, and Sivakamavalli Jeyachandran**

## 7.1 Introduction

Shrimp aquaculture has seen immense growth over the past three decades to become a major global food industry which serves the seafood sector based on consumer demand and plays a major role in the economic development of various poor coastal aquaculture communities across the globe. The environmental changes may also serve as a pathway or precondition for the emergence and spread of the disease. Shrimps are taken from their wild environments and fed with artificially composed or synthetic pellet feeds. For higher yield and production, shrimps are stocked in areas with high population density (Ashraf et al. 2011). Due to the changes in water quality, shrimps are exposed to stress and are transported overseas, either as live or as frozen products for consumers. These practices have provided opportunities for increased disease outbreaks due to stress and environmental conditions, exposure of new pathogens to the culture pond, and the rapid transmission of infectious pathogens and the spread of disease. Several scientific domains are participating in this sector to eradicate or control the viral pathogenesis mechanism in the aquaculture industry. Among these, nanotechnology can be used as an efficient method to eradicate the devastating disease-causing pathogenic viruses (WSSV, YSV, and TSV), which collapse the socio-economic value of a country. Nanotechnology is defined as the study and application of functional material devices and systems through the control of matter at the atomic and molecular levels, such as at the nanometer scale,

H. Chellapandian · S. Jeyachandran (✉)
Department of Orthodontics, Saveetha Dental College and Hospital, Chennai, Tamil Nadu, India

M. S. Aadham · M. Jaabir
PG & Research Department of Biotechnology and Microbiology, National College (Autonomous), Tiruchirappalli, Tamil Nadu, India

 
V. K. Arivarasan et al. (eds.), *Nanotechnological Approaches to the Advancement of Innovations in Aquaculture*, Nanotechnology in the Life Sciences, https://doi.org/10.1007/978-3-031-15519-2_7

and the novel phenomena and properties of matter are exploited at that nanometer scale. There is no comprehensive and concise definition of nanomaterials and nanoparticles. So far, there are various definitions that have been proposed by the government, industry researchers, even private labs, and other research institutes and organizations (Arulmoorthy et al. 2020). The definition of nano is still under debate because there is no proper definition for nano, which is basically said to be based on size. These kinds of nano-based functional devices are used in all the fields of science, such as chemistry, physics, and biology. The whole world is made up of atomic particles, so the effect of nanoscience plays a vital role in bioremediation and bio restoration in aquaculture. Here we are going to discuss the viral diseases that have majorly affected the shrimp culture industry over the past two decades. Some of the diseases are addressed in Table 7.1.

## 7.2 White Spot Syndrome

White spot syndrome virus is a type of virus within the genus *Whispovirus* in the Nimaviridae family. Basically, WSSV virions are rod- or elliptical-shaped, symmetrical, and approximately 80–120 nm in diameter and 250–380 nm in length as per measurement. It is noted that there are 39 structural proteins at the end of the flagellum like appendages on virions. The virulence of the WSSV shows differences among diverse isolate varieties, and it has a different-sized circular double-stranded DNA genome. WSSV, formerly known as systemic mesodermal and ectodermal baculovirus, is a non-occluded baculovirus-like agent infecting many penaeid species, first examined in *P. japonicus*. Within 2–5 days of time, WSSV has the ability to cause up to 100% mortality in commercial shrimp farms, and as a result, the shrimp aquaculture industry faces a huge economic loss across the globe. WSSV mostly attacks shrimps at the post-larval stage. However, disease occurrence is even seen in growing juvenile shrimp of all ages and sizes, but mostly from 1 to 3 months after stocking in grow-out ponds. WSSV outbreak can occur irrespective of farming system, stocking density, water quality, and salinity. The subsequent source of WSSV infection is dead or infected shrimps; cannibalism is said to be one of the main reasons for the transmission of the virus. Even infected pond water acts as the waterborne route (Sivakamavalli and Vaseeharan, 2014). In shrimp farms, the major source of infection is due to the infected spawners and postlarvae (PL) from the infected broodstocks of hatcheries. The vertical transmission of WSSV by trans-ovum also acts as a major reason for the transmission of the disease. Rotifers, marine molluscs, polychaete worms, *Artemia salina*, Ephydridae insect larvae, and sea slaters (Isopoda) are the major disease-carrying vectors of the white spot syndrome virus.

## 7.3 Symptoms

- In moribund shrimp, gross signs are seen along with rapid mortality.
- Lethargy, black gills, anorexia, broken antennae.

**Table 7.1** Common viral diseases and its causative agents in different shrimp species

| Disease | Virus | Abbreviation | Genome | Family | Genus | Species affected |
|---|---|---|---|---|---|---|
| White spot syndrome | White spot syndrome virus | WSSV | dsDNA | Nimaviridae | *Whispovirus* | All farmed marine penaeid shrimp species |
| Infectious myonecrosis | Infectious myonecrosis virus | IMNV | dsRNA | Totiviridae | *Giardiavirus* | *P. vannamei* |
| White tail disease | Macrobrachium rosenbergii nodavirus | MRNV | (+) ssRNA | Nodaviridae | Related to *Alphanodavirus* and *Betanodavirus* | *P. vannamei* |
| Infectious hypodermal and hematopoietic necrosis | Infectious hypodermal and hematopoietic necrosis virus | IHHNV | ssDNA | Parvoviridae | *Brevidensovirus* | *P. monodon*<br>*P. vannamei*<br>*P. stylirostris* |
| Taura syndrome | Taura syndrome virus | TSV | (+) ssRNA | Discistro-viridae | *Cripavirus* | *Penaeus vannamei* |

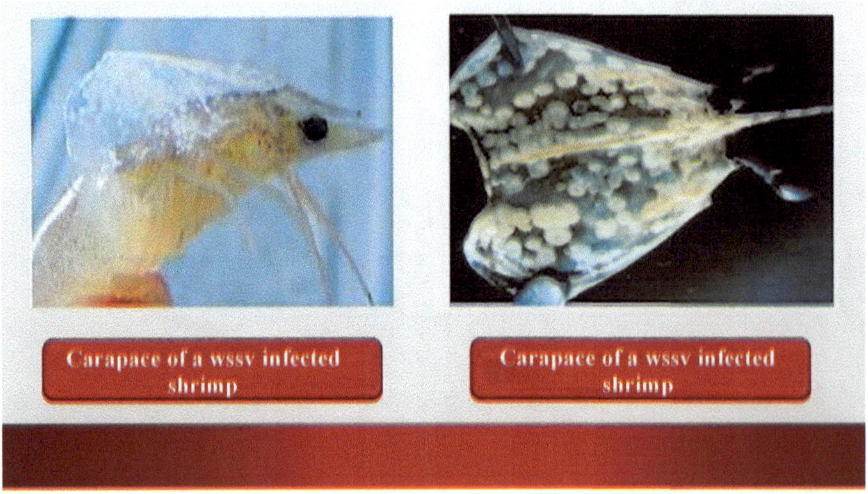


**Fig. 7.1** WSSV-infected carapace of a shrimp

- Swollen branchiostegites.
- On the cuticle of the cephalothorax or carapace, white deposits of calcium can be seen.
- Display expanded chromatophores and cephalothorax cuticles.
- Along with white spots all over the body surface (Fig. 7.1).

The target tissues or cells for the white spot syndrome virus are of mesodermal and ectodermal origin: the cuticular epithelium, lymphoid organs, the heart, haematopoietic tissue, the cuticular epidermis, subcuticular tissue and other connective tissue, the antennal gland, and the hindgut. In chronic cases, the nervous system and compound eyes are infected, and the enlarged haemal sinuses and interstitial spaces are caused by the diseased lymphoid organ of shrimp, which will result in a hypertrophied yellowish hepatopancreas. Sometimes an empty gut also indicates the disease occurrence, along with cuticular epiboint fouling and lymphoid organ swelling. Marine penaeid shrimps are the major victims for white spot syndrome virus which also includes marine, brackish water and freshwater prawns, crayfish, crabs, shrimps and lobsters. *Macrobrachium rosenbergii*, *Penaeus monodon*, *Litopenaeus vannamei*, *L. stylirostris*, *M. japonicus*, *F. indicus*, *Procambarus clarkia*, and *P. leniusculus* are some of the common examples of farmed aquatic animals affected by WSSV.

## 7.4 Control and Prevention

- There have been no vaccines developed against WSSV. In recent studies, nanotechnology has a really good effect when compared to other methodologies.

- The pathogenicity of WSSV is decreased at low temperatures (12 ± 2 °C), and also it inhibits the mortality rate in crayfish and shrimp.
- Certain bio-components, such as polysaccharides bound with sulfate, and micro-algae cell walls, can be used as immunostimulants against WSSV in shrimp.
- It can be prevented or avoided by stocking season-specific pathogen-free (SPF) larvae or PCR-tested larvae in the culture system.
- In *P. monodon*, the usage of β-1,3-glucan as a feed mix along with shrimp feed effectively improves the immunity and survival.
- Certain probiotics of *Bacillus* species and *Dunaliella* extracts show some reasonable resistance in shrimp against WSSV infection.
- The epizootics can be reduced by good farming and farm management practices, along with the control of environmental variables such as PH, temperature, salinity, and other factors, which play a vital role in the genetic improvement of the penaeid species of shrimp.

  Several approaches might be undertaken through molecular biology approaches, but at the same time, the rapid development of nanotechniques is an unavoidable arena in every platform to eradicate and control scientific problems. On this connection, various nanoparticles, nanocomposites, nanospheres, nanorods, and nanocubes have been applied in bacterial inhibitory mechanisms in agriculture as well as aquaculture platforms. However, few viral infections wouldn't be possible to eradicate completely from the culturing sectors, and at the same time, we can control the viral infections through different nanoformulations.

## 7.5 Infectious Myonecrosis

Penaeid shrimps are the main victims of the infectious myonecrosis virus (IMNV). So far, the IMN has been reported only in shrimps from rearing ponds, but not in the wild populations due to the lower density of shrimps in the marine environment. *Litopenaeus vannamei* (Pacific whiteleg shrimp) and *Farfantepenaeus subtilis* (Southern brown shrimp) are easily prone to the IMNV, and this makes a huge economic loss for a country in the food sector. In *P. monodon*, due to IMN infection, there have been no reports of death; even though there are no death reports for *P. monodon*, it can act as a potential carrier of the virus. When compared to other virulent shrimp viruses, the IMNV does not affect any vital organs because its target tissue is skeletal muscles, so, therefore, IMNV is not that fatal. The damaged muscle tissues can be repaired within the early five stages of infection (Tang and Kotov 2005). Various studies have been done to understand the source of potential carriers of the IMNV and (da Silva et al. 2015) reported that a zooplankton named *Artemia salina*, which is used as a live feed, acts as the major carrier of IMNV infection. But the ones which fed on infected artemia do not show any mass mortality. This may be because they act as a source of IMNV only in grow-out culture ponds. Certain worm types such as polychaete worms and blood worms showed a positive response to IMNV infection. The presence of the virus may be due to the ingestion of contaminated tissues or water from the infected area. The IMNV has some peculiar and specific characteristics. These specific characteristics and behaviors made this double-stranded RNA virus one of the most dangerous shrimp industry viruses because it is a stress-dependent virus.

## 7.6 White Tail Disease

The most extensively cultivated shrimp species was *Litopenaeus vannamei* (*L. vannamei*) across the coastal communities of the world. This is because of its high yield and high market value. The emergence of various viral diseases such as WSSV, YHD, TSV, and alpha- and betanodavirus is due to the continual expansion and intensification of shrimp farming. Among the abovementioned diseases, IMNV and alpha- and betanodavirus are the major causative pathogens for the emergence of white tail disease (WTD), as well as *Macrobrachium rosenbergii* nodavirus (MrNV) identified in freshwater prawns (Naveenkumar et al. 2020). The target tissue for both of the viruses are primarily skeletal muscles, which results in gross signs such as white or opaque tail and histopathologically shows muscle necrosis and changes in lymphoid organs as spheroids, moreover in all Penaeid shrimps. In recent studies, it has been identified that a highly virulent *Vibrio harveyi* strain **HLB0905** is the causative agent for the white tail disease. This can be examined through microscopical morphology analysis, genome sequence analysis, and bacterial isolation. Interestingly, in *L. vannamei*, the particular strain of *V. harveyi* not only causes "white tail disease" (WTD) but also was nonluminescent in nature (Zhou et al. 2012).

## 7.7 Infectious Hypodermal and Hematopoietic Necrosis

This is a viral disease which mainly infects the penaeid shrimp variety and may cause up to 90% of mass mortality in culture ponds as well as in wild environments in the Pacific region. Infectious hypodermal and hematopoietic necrosis virus (IHHNV) can be seen in the population of *Penaeus stylirostris* and in *Litopenaeus vannamei*. Recent studies made by the shrimp-farming industries and the researchers of aquaculture have developed several broodstocks which are resistant to the IHHNV in both *P. stylirostris* and *L. vannamei*. *Decapod pestylhamaparvovirus* is the causative pathogen of IHHN, which is notably a single-stranded DNA virus and also the smallest known virus at 22 nm that infects shrimp. The virus has been classified as *P. stylirostris hamaparvovirus* (Natrah et al. 2011). The replication of the virus is through the host cell cytoplasm, so it can be even classified as a picornavirus. The symptoms of IHHN infection are reduced food consumption, increased mortality rate of 80–90% in the juveniles grown in high-density and raceway cultures, and even cannibalism in culture systems of penaeid shrimp. Some of the species of Penaeidae family survive IHHNV infection, but they carry the virus for their whole life and transfer it to the progeny and other populations by vertical or horizontal transmission. IHHNV infection is also known as runt deformity syndrome (RDS) in *L. vannamei*.

## 7.8 Taura Syndrome

Taura syndrome virus (TSV) is a virus which acts as the causative pathogen for taura syndrome (TS) infection. *Litopenaeus vannamei* and *P. stylirostris* are the common host species of the TS virus and are easily susceptible to TS infection, whereas TS can be classified into three distinct phases: acute, transitional, and chronic. In the acute phase, a histological section of epithelium and cuticular epithelium shows lesions. But in the transition and chronic phases of the disease, there are no signs of lesions in the epithelium, and the TSV infection can be detected only by molecular and antibody-based methods. By good farming practices, an individual shrimp can be protected from chronic TSV infections which may persist for the whole lifespan. The cuticular epithelium of the exoskeleton, the gut region, the gills, and the appendages are the major sites of TS virus replication, and initially, it infects these regions. The histological sections of gastroenteric organs, cardiac and striated muscles, the ventral nerve cord, and its branches and ganglia do not show any signs of TSV infection and are usually negative for TSV by ISH (in situ hybridization). Pale reddish discoloration on the tail fan and pleopods is seen in the penaeid shrimp, postlarvae, or older shrimps of *Penaeus vannamei*, which is also known as red tail disease. The expansion of the red chromatophores in the cuticular epithelium is the reason for the color change and the peppered appearance. The reddish discoloration of the tail fan and uropods has rough edges on the cuticular epithelium. These are the symptoms of the acute phase of Taura syndrome, which shows lesions.

## 7.9 Yellow Head Disease (YHD)

The causative pathogen for Yellow head disease (YHD) is Yellow head virus (YHV). YHV is an ssRNA virus which is closely associated with the families Coronaviridae and Arteriviridae. Gill-associated virus (GAV) is closely related to the YH virus, which comes under the genus Okavirus and has a wide range of hosts. Naturally, the infections occur particularly in *P. monodon*, but infections are also reported in *P. japonicus*, *L. vannamei*, *P. setiferus*, and *P. stylirostris. Penaeus merguiensis* and *Palaemon styliferus* are resistant to YH disease, but they act as carriers of viral pathogens. Euphausia spp. (krill), *Acetes* spp., and other small shrimps are also reported to act as carriers of YH viruses. YHD was reported in Thailand for the whole year of 1999. Aggregation of diseased shrimps can be seen at the very edge of the pond. Diagnosis of Yellow head disease can be confirmed by the following symptoms: discoloration of the hepatopancreas, which results in the yellowish appearance of the cephalothorax; this is the reason for the name of the disease. The overall appearance of the shrimp becomes abnormally paler than usual, and it affects many tissues such as gills, lymphoid organs, hemocytes, and connective tissue. Histopathology: degenerative changes in nuclei and presence of cytoplasmic basophilic inclusion bodies. Older shrimp and postlarvae (PL) of 15–25 days are infected, while postlarvae of more than 15 days appear resistant. Mass mortalities of up to 100% are observed within 3–5 days (Hasson et al. 1995).

## 7.10 Nanotechnology in Detection and Disease Management in Shrimp

Nanoarrays, protein arrays, and nanopores are used in the detection of viruses. These are nanomaterials which can be used in the management of viral diseases in shrimp. Many researchers are trying to use nanoparticles as nano-based sensors and nano-vaccines to prevent disease outbreaks and run mortality in growing-out ponds (Flegel, 2019). A simple and rapid immune-chromatographic test-strip kit used for WSSV detection is being developed by various researchers and organizations with gold nanoparticles (AuNPs) conjugated with monoclonal antibody W29 as the detector antibody (Khosravi-Katuli et al. 2017). In this method, rabbit anti-recombinant protein antibody is used in permutation with W28 monoclonal antibody as an arrest complex in the test line and goat anti-mouse IgG antibody (GAM) as an arrest antibody in the control line. This method is demonstrated using chromatography through the nitrocellulose membrane. While the test is four times less sensitive than the dot blot technique and is much less sensitive than single-step PCR, it was suggested to test samples from the individual as well as combined shrimp samples to validate high probabilities of WSSV infection and disease outbreaks. The working mechanism of this kit is simple, easily approachable, and very convenient for obtaining immediate results without the help of hi-fi tools or technicians with previously learned skills (Rather et al. 2011).

A rapid and label-free detection of the WSSV with a surface plasmon resonance (SPR) device based on gold films equipped with electroless plating was developed by Lei et al. (2008). The electroless-plated gold films were modified with single chain fragment variable (scFv) antibody molecules and were used in combination with a highly sensitive SPR binding between a WSSV sample and the anti-WSSV scFv antibody pre-immobilized against the sensor surface device for detection of the virus present in different concentrations in shrimp hemolymph matrix. The method could detect the application of WSSV as low as 2.5 ng/mL in shrimp hemolymph, so enzyme-linked immunosorbent assays cannot be done due to the lower CFU levels (Govindaraju et al. 2019). Though time-consuming, the method is cost-effective and highly reproducible, which could help increase sensors performance-price ratio of SPR sensors, suggesting that it can be used widely in the fabrication of SPR sensor substrates. To develop a prototype for the development of a simple and cheap diagnostic tool for in-field testing in farms, Thiruppathiraja et al. (2011) used alkaline phosphatase along with AuNPs, which is used as a primary vaccine for WSSV infection. Also they have successfully enhanced the immunity through conventional methods, where visually it can be seen up to 1 ng/mL of purified WSSV (Thiruppathiraja et al. 2011). However, though protein-based immunodetection methods are easier to perform, they lack the sensitivity appropriate to adequate signal amplification, which is a major impediment to taking this technique forward for wider application. Several genetic diseases can be treated by the development of novel carrier systems for gene delivery; this can be done by nanotechnology by means of nanoparticles. Certain approaches utilize DNA complexes containing proteins, lipids, polypeptide carriers, as well as nanoparticles with ligands that are capable of targeting specific DNA complexes to the cell-surface receptors on target cell sites and directing intracellular trafficking and delivery of DNA towards the nucleus. One among those is the formation of complexes between chitosan and DNA polyplexes for oral gene therapy applications. Nanoparticles loaded with plasmid DNA could serve as a proficient gene delivery system due to their quick movement to the cytoplasmic compartment from the degradative endo-lysosomal compartment. Repeated gene expressions can be seen due to the constant discharge rate of DNA. Studies on the protective efficacy of the oral delivery of a DNA construct containing the VP28 gene of WSSV encapsulated in chitosan nanoparticles in P. monodon showed 50% survival even after 30 days, while the control registered 100% mortality (Govindaraju et al. 2019).

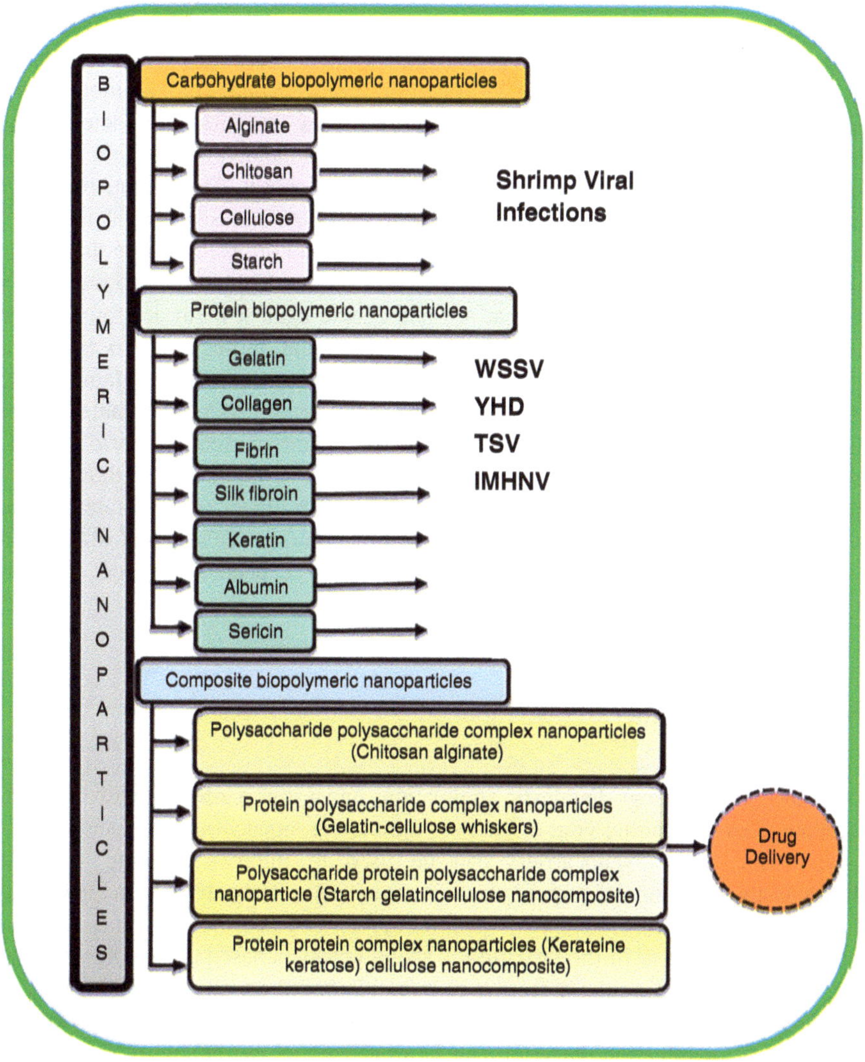

## 7.11 Using AgNPs Against Shrimp Viral Diseases

It is proven that various types of pathogens infecting humans can be treated with nanoparticles, especially AgNPs, but still the effects of AgNPs on shrimp diseases have limited information (Boverhof et al. 2015). The source of evidences from in vivo studies shows that AgNPs have the potential for antimicrobial activity against certain shrimp diseases like WSSV and bacterial diseases caused by *Vibrio parahemolyticus* and other *Vibrio* species. WSSV is a vigorous strain of virus which rapidly spreads over the culture area and causes a complete spread of white spot disease on the shrimp ponds in a particular area, which makes it an alarming danger for farms around the globe. Vibriosis, brown spot disease, and acute hepatopancreatic necrosis syndrome (AHPNS) are some of the bacterial diseases caused by *Vibrio*. These bacterial types are ubiquitous in all coastal and brackish water ecosystems in all parts of the world. The abovementioned disease causes considerably huge economic losses in the seafood sector in all the coastal communities, farms, and hatcheries across the globe. Recent studies show that viral and bacterial diseases of shrimp can be rectified by the application of nanoparticles, these applications can be used for therapeutic and prophylactic purposes. The administration of drugs such as vaccines and other medications before the disease outbreak is known as a prophylactic application, whereas the application of drugs for an ongoing active disease in growing-out shrimp ponds is known as a therapeutic application. In marine shrimps, *vibrio* species that are pathogenic against them are found to be inhibited by the AgNPs synthesized through chemical, biological, and even biochemical methods (Shaalan et al. 2016). In Penaeidae shrimp family like *Fenneropenaeus indicus, P. monodon*, and *L. vannamei*, it is proven that to some extent nanoparticles prevent certain diseases of shrimp caused by *Vibrio* species, which include vibriosis and APHND in postlarvae of *L. vannamei*. In those cases, the prophylactic effects of nanoparticles translate to a higher survival rate of individuals exposed to bacteria after AgNPs administration over a period of time, in contrast to untreated animals that presented the expected elevated mortalities (Camacho-Jiménez et al. 2020).

## 7.12 Conclusion

Nanotechnology brings diagnostic applications on various platforms, with diagnostic approaches including viral diseases in agriculture, aquaculture, and shrimp disease management. Nanocarriers can be used to improve stability, increase immunostability, and also target specific immune cells. They are used in the formulation of antigens for delivery to specific target sites, and there are lots of other adjuvant advantages. In the near future, many products based on nanotechnology applications, including the detection of infection based on nanomaterials and disease management, will be on field trials, and promising results have been seen so far. Disease or infection detection can be done by the abovementioned products, which can be rolled out in the market for field trials. Even treatment and prevention of viral pathogens such as WSSV, IMNV, IHHNV, YHD, and TSV can be done using nanotechnology and its nanomaterials. The rapid development of nanotechnology in aquaculture, especially in the shrimp industry, can revolutionize the diagnosis and treatment of shrimp disease. Nowadays, nanosensors in the pond environment are used to detect variations in water, such as mineral deficiency, mortality, ammonia (sludge) levels, and even malicious disease outbreaks. To prevent environmental and disease stresses, nano-based functional feeds are developed. Several other ideologies include the removal of pollutants from aquaculture pond systems and the enhancement of finfish and shellfish growth for sustainable aquaculture farming practices.

## References


Arulmoorthy MP, Anandajothi E, Vasudevan S, Suresh E (2020) Major viral diseases in culturable penaeid shrimps: a review. Aquac Int 28(5):1939–1967

Ashraf M, Aklakur M, Sharma R, Ahmad Sh KM (2011) Nanotechnology as a novel tool in fisheries and aquaculture development: a review. Iran J Energy Environ 2(3):258–261

Boverhof DR, Bramante CM, Butala JH, Clancy SF, Lafranconi M, West J, Gordon SC (2015) Comparative assessment of nanomaterial definitions and safety evaluation considerations. Regul Toxicol Pharmacol 73(1):137–150

Camacho-Jiménez L, Álvarez-Sánchez AR, Mejía-Ruíz CH (2020) Silver nanoparticles (AgNPs) as antimicrobials in marine shrimp farming: a review. Aquac Rep 18:100512

da Silva SM, Lavander HD, de Santana Luna MM, Gálvez AO, Coimbra MR (2015) Artemia franciscana as a vector for infectious myonecrosis virus (IMNV) to Litopenaeus vannamei juvenile. J Invertebr Pathol 126:1–5

Flegel TW (2019) A future vision for disease control in shrimp aquaculture. J World Aquacult Soc 50(2):249–266

Govindaraju K, Itroutwar PD, Veeramani V, Kumar TA, Tamilselvan S (2019) Application of nanotechnology in diagnosis and disease management of white spot syndrome virus (WSSV) in aquaculture. J Clust Sci 15:1–9

Hasson KW, Lightner DV, Poulos BT, Redman RM, White BL, Brock JA, Bonami JR (1995) Taura syndrome in Penaeus vannamei: demonstration of a viral etiology. Dis Aquat Org 23(2):115–126

Khosravi-Katuli K, Prato E, Lofrano G, Guida M, Vale G, Libralato G (2017) Effects of nanoparticles in species of aquaculture interest. Environ Sci Pollut Res 24(21):17326–17346

Lei K, Li F, Zhang M, Yang H, Luo T, Xu X (2008) Difference between hemocyanin subunits from shrimp Penaeus japonicus in anti-WSSV defense. Dev Comp Immunol 32(7):808–813

Natrah FM, Ruwandeepika HD, Pawar S, Karunasagar I, Sorgeloos P, Bossier P, Defoirdt T (2011) Regulation of virulence factors by quorum sensing in Vibrio harveyi. Vet Microbiol 154(1–2):124–129

NaveenKumar S, Rai P, Karunasagar I, Karunasagar I (2020) Recombinant viral proteins delivered orally through inactivated bacterial cells induce protection in Macrobrachium rosenbergii (de Man) against White Tail Disease. J Fish Dis 44(5):601–612

Rather MA, Sharma R, Aklakur M, Ahmad S, Kumar N, Khan M, Ramya VL (2011) Nanotechnology: a novel tool for aquaculture and fisheries development. A prospective mini-review. Fish Aquacult J 16:1–5

Shaalan M, Saleh M, El-Mahdy M, El-Matbouli M (2016) Recent progress in applications of nanoparticles in fish medicine: a review. Nanomedicine 12(3):701–710

Sivakamavalli J, Vaseeharan B (2014) Variations in biochemical and histological characteristics of WSSV infected green tiger shrimp Penaeus semisulcatus. J Recept Signal Transduct 34(5):386–395

Tang Z, Kotov NA (2005) One-dimensional assemblies of nanoparticles: preparation, properties, and promise. Adv Mater 17(8):951–962

Thiruppathiraja C, Kumar S, Murugan V, Adaikkappan P, Sankaran K, Alagar M (2011) An enhanced immuno-dot blot assay for the detection of white spot syndrome virus in shrimp using antibody conjugated gold nanoparticles probe. Aquaculture 318(3–4):262–267

Zhou J, Fang W, Yang X, Zhou S, Hu L, Li X, Qi X, Su H, Xie L (2012) A nonluminescent and highly virulent Vibrio harveyi strain is associated with "bacterial white tail disease" of Litopenaeus vannamei shrimp. PLoS One 7(2):e29961

# Chapter 8
# Applications of Nanoparticles in Aquaculture

**Prachi Vibhute, Mohammed Jaabir, and Jeyachandran Sivakamavalli**

## 8.1 Introduction

Protein demand is increasing as the world's population and economy grow at a rapid pace. Aquatic protein resources are widely valued due to positive health impacts and key food composition aspects, and as a result, global aquaculture has risen at an astonishing rate recently. Aquaculture currently provides 50% of the world's fish for human consumption. Global fish production increased by approximately 171 million tonnes in 2016, with a total first-sale value of US$ 362 billion for fisheries and aquaculture industry. Food fish consumption per capita increased from 9.0 kg in 1961 to 20.2 kg in 2015, an average yearly increase of around 1.5%, and was expected to increase to almost 20.3 and 20.5 kg in 2016 and 2017, respectively (FAO 2018). Nonetheless, as the sector develops, there is still a question mark over its long-term viability, as the effect of ever-increasing aquaculture waste has a negative influence on both productivity within aquaculture systems and the ambient aquatic ecology. There is a significant gap in technological advancement for drug use, treatment approaches, water quality management, the development of tailored fish for good health, productivity driven by epigenetic as well as nutrigenomic interactions, better breeding success through efficient delivery of maturation as well as spawning inducing agents, nutraceutical delivery for fast growth promotion, and reduced culture time. To address these obstacles, a combination approach of

P. Vibhute · M. Jaabir
PG & Research Department of Biotechnology & Microbiology, National College, Tiruchirappalli, Tamil Nadu, India

J. Sivakamavalli (✉)
Department of Orthodontics, Saveetha Dental College and Hospitals, Tamil Nadu, Chennai, India

 
V. K. Arivarasan et al. (eds.), *Nanotechnological Approaches to the Advancement of Innovations in Aquaculture*, Nanotechnology in the Life Sciences, https://doi.org/10.1007/978-3-031-15519-2_8

comprehending, integrating, engaging, and deploying new science and technological techniques in order to sustain desirable aquaculture is required (Aklakur et al. 2016).

Nanotechnology is being used to deliver nutrients, vaccines, and other biological elements to various fish systems (Shaw and Handy 2011; Handy et al. 2008a). Nanoparticles have been used in a variety of medical applications, including diagnostics, immunisation, and medication and gene delivery. There are a variety of nanoparticles that are less harmful to the various cells in the fish immune system. When it comes to transmitting biological components, nanoparticles are less expensive than other materials. As a result, nanoparticles can be utilised in large-scale experiments in animal systems to research various aspects of food intake via various carriers (Hosseini et al. 2016; Vimal et al. 2012).

Because some biological materials have low specificity, they can't strongly bind to pathogens and thus can't be employed as a primary target for viruses and bacteria (Yue et al. 2017). It leads to breakthroughs in nanotechnology, which has transformed the generation of biological materials. In aquaculture, nanomaterials have piqued interest as a particular and sensitive method for diagnosing bacterial, fungal, and viral infections (Sarkar et al. 2015). Cu-based nanoparticles, Ag nanoparticles, metal oxide nanoparticles such as ZnO and $TiO_2$ NPs, and composites of various metals are all examples of nanoparticles (Khan et al. 2015). These NPs are less harmful than their biological materials in comparison to nutrients and drug delivery in fish, although metal toxicity in fish can be examined in a better method to compare and contrast the effects of nonmetals on the different cells of fish. These various types of NPs can be engineered to serve an important role in disease diagnosis and fish treatment. Each nanoparticle has its own distinct properties that are developed for medication delivery and other medical applications. Silver-based nanoparticles are the most prevalent type of nanoparticle used in illness detection in fish and other animals (George et al. 2012).

Disease is a major factor in fish mortality, especially in young fish. Pathogenic illnesses and nonpathogenic diseases are the two types of fish diseases based on their infectiousness. Noninfectious diseases include gas bubble diseases caused by increased aeration, nutritional illnesses caused by a lack of certain nutrients (such as vitamins and minerals), pollution-related disorders (industrial and agricultural), and neoplastic and hereditary abnormalities, which refer to abnormal cell growth in any organ that causes the organ to start losing its function and structure. The infectious disease, on the other hand, is regarded as a particularly deadly condition since it is transmitted from one fish to another and results in massive mortality. Infectious diseases are divided into three categories: bacterial, fungal, and parasitic (Ali et al. 2014, 2018). Control initiatives have been adopted over the years for a large number of chemicals that are presently being utilised in the establishment of the aquaculture industry. Hormones, vitamins, antibiotics, and a few other compounds were tested in aquaculture systems for various cures. Although they have favourable advantages, they cannot be recommended due to their long-term and other negative effects. Synthetic therapies include disinfectants (e.g. hydrogen peroxide and green malachite), antibiotics (e.g. sulfonamides and tetracyclines), and anthelmintics (e.g.

pyrethroid insecticides and avermectins); neutral remedies include probiotics and essential oils (Citarasu 2010; Rawn et al. 2009). Nanotechnology is thought to offer a solution for preventing and monitoring illnesses and infections, as well as multiplying the benefits of aquaculture. Antibacterial and antifungal surfaces made utilising porous nanostructures, nanosensors in aquaculture environments for pathogen detection in water, and nanodelivery of veterinarian goods and fish medicines via fish meals are some of the fish health applications of nanotechnology. Nano-trace element usage is up to 100 times greater than standard inorganic trace element usage, which is often extremely limited because the former enters the animal body by direct penetration (Nasr-Eldahan et al. 2021; Shah and Mraz 2020).

Because nanomaterials work on the very same level as a particle-infecting virus or diseases, controlling viruses, bacteria, and fungi would require early detection and removal of pathogens (Nasr-Eldahan et al. 2021; Shah and Mraz 2020; Luis et al. 2019). Because nanoencapsulated materials are robust and can withstand high temperatures or acidity, chitosan-based wrapping around vaccinations as a nanoencapsulation carrier for effective therapy of bacteria and viruses causing fish sickness has been documented. These nanomaterials are proving to be highly advantageous in the growth of pathogen-free fish seedlings and shrimp or prawn post-larvae in agriculture (Nasr-Eldahan et al. 2021).

## 8.2 Nanoparticles

Materials with one or more dimensions in the nanometre (1–100 nm) range are called nanoparticles. Materials in the nanoscale range have been linked to unique features and applications that are distinct from their respective bulk counterparts. Nanoparticles have a large exposed surface area per unit volume due to their small size, which increases their chemical reactivity. It's also said to be particularly stable in high-temperature and high-pressure environments. These are easily absorbed by animals' gastrointestinal tracts, making them more effective at lower doses than bulk materials (Chris et al. 2018). They also interact more efficiently with organic and inorganic compounds within animal bodies, and they can cross the small intestine to reach the blood, brain, liver, and other organs (Hillyer and Albrecht 2001). For particle size homogeneity and eco-friendliness, nanomaterials for feeding animals are best synthesised via chemical or biological techniques (Iravani and Zolfaghari 2013). Nanometals (e.g. silver nanoparticles (NPs)), metal oxides (e.g. $TiO2$ NPs), carbon-based materials such as carbon nanotubes (CNTs), and carbon spheres (e.g. C60 fullerenes, dubbed "buckyballs" by the media), and composites made of several substances such as nanoceramics and quantum dots have all been produced in various chemical forms (Boxall et al. 2007; Stone et al. 2010). Combinations of known hazardous metals, such as ZnS and CdSe, employed in quantum dots for novel light-emitting diodes can be used in the latter (Bae et al. 2009). NMs, on the other hand, have the possibility of an unlimited number of chemistries. Hydroxyl groups, carboxylic acid groups, sulphate moieties, and other functional groups can be applied to the surface of NMs. Compounds on the NM's surfaces can be covalently bounded and so fundamentally part of the material's composition, or they can be a loosely attached surface coating (e.g. citrate-coated metal NPs). "Second-generation" NMs are beginning to appear, with sophisticated three-dimensional structures and/or several chemical components. For very specialised purposes, NMs are already being made with highly functionalised surfaces (e.g. functionalised CNT) (Handy et al. 2008b; Ju-Nam and Lead 2008).

Nanoparticles can be made in a variety of ways, including chemical, biological, and physical techniques. The biological method is one of the most important strategies for drug delivery and other biological material supplementation across the cell membrane. Because of the low cost and great efficiency of the synthesis of nanoparticles, this technology has a significant advantage over chemical and physical methods. The attention of other materials and their incorporation in different portions of the cell are increased by the green creation of nanoparticles. Traditional methods, on the other hand, are unreliable for the preparation of NPs because they require a huge number of reagents and chemicals for the large-scale production of nanoparticles, both at the industrial and commercial levels. As a result, old methods are no longer viable due to low demand and expensive costs when compared to technological advancements in nanotechnology. Nanoparticles can undergo several modifications during synthesis and modification before being transformed into their final form, whereas other materials, other than nanoparticles, require a large number of raw components and are time-demanding (Munawar 2021).

Fish are heterotrophic organisms with numerous organs and tissues. Fish can ingest algae or other heterotrophs, depending on the species. Various nanoparticles can be employed or carried to the various cells of the fish in order to provide the appropriate treatment at the appropriate location. Because of its unique action, once the nanoparticle reaches its designated location, it will begin to perform its activities at the cellular and molecular levels. The gill allows for gas exchange between both the organism's internal and external aqueous environments. Other chemicals, such as metal nanoparticles and organic compounds, can interact with fish gill cells and eventually get into the bloodstream during this exchange process (Kok et al. 2020).

Supplementing fish diets with nanoparticles of elements such as selenium, iron, and other sources could help them develop faster. The method can be used to lower the cost of water treatment in aquariums and commercial fish ponds. Nanotechnologies, according to researchers, have the potential to create disease-free and pollution-free fishponds. Another potential application for nanotechnologies is the use of various conservation and packaging techniques to ensure seafood safety by preventing mildew and microbial decomposition (Can et al. 2011).

Nanoparticles also revealed that various cells have distinct reactions to changes in the rate of fish death. Many nanoparticles can penetrate antigen-presenting cells via various routes and elicit suitable immune responses to the antigen. It also depends on the quality of each nanoparticle that is designed to execute a certain role in a cell. Biodegradable polymers, nanoliposomes, carbon nanotubes, calcium phosphate, and immunostimulating complexes (ISCOMs) are among the nanoparticles utilised in fish vaccine administration. Poly (lactic-co-glycolic acid) and chitosan are the most investigated forms of nanoparticles to date (Pandey and Prajapati 2018; Kumar et al. 2018; Dawood et al. 2020).

## 8.3 Action of Nanoparticles

Concerns about nanosafety still exist, and they must be addressed before full-scale implementation. Nanoparticle toxicity is determined by a complex interaction of particle parameters such as diameter, form, surface charge, concentration, time of exposure, nature of the nanoparticles, medium composition, route of particle administration, and the immune system of the target species. Despite the current data, various criticisms obstruct a complete knowledge of the safety of nanoparticles in aquaculture. To begin with, the manner in which nanoparticles are administered in aquaculture might vary greatly: they can be added to food, water media, or aquaculture infrastructure. Nonetheless, present aquatic toxicology studies are insufficient to meet the demand for nanoparticle safety in aquaculture, such as administration route, concentration, and exposure time. Concentrations are sometimes lower or greater than those used in aquaculture or expected to be used, resulting in unrealistic outcomes. As a result, inferring the potential negative impacts on the final consumer is impossible. It is vital to investigate the safety of nano-based aquaculture, taking into account not only relatively short treatment periods (less than 40 days) but also the entire cultured product life cycle, including water quality, from the egg/larva to the table. Second, because aquatic organisms are maintained in diverse conditions, nano-based goods can behave quite differently in terms of the generated consequences; hence, it might be interesting to investigate how environmental parameters such as pH, salinity, and temperature influence nanosafety (Khosravi-Katuli et al. 2017).

### *8.3.1 Selenium (Se)*

Selenium is a trace element that is necessary for normal bodily processes and animal metabolism (Prashanth et al. 2015). It has a substantial impact on the physiology of fish by improving the animal's physiological and immunological systems. Selenium supplementation protects cells from harm and is important for fish development, fertilisation, and immunological function. It also protects the organism from oxygen-free radicals, which are formed in stressful situations or when an animal is exposed to certain types of toxicity (Khurana et al. 2019). As per Le et al. (2014), selenium supplementation improves fish immunity by increasing lysozyme activity as well as fish red blood cell (RBC) count, with RBC and hematocrit proportions in tilapia being improved with the right diet (El-Hammady et al. 2007). The nano form of selenium has the most favourable effect, as it is more effective than the bulky version. The nano form of selenium is a unique type that draws more interest than inorganic and organic forms, wherein inorganic compounds are more poisonous than organic compounds, owing to their large bioavailability and lower toxicity (Khurana et al. 2019; Shi et al. 2011). The biological features of selenium nanoparticles (SeNPs) are dependent on their size: tiny particles have more activity (Torres et al. 2012). In comparison to other organic and inorganic oxidation states, nano-selenium (nano-Se) advantages from the capacity to utilise selenium at zero oxidation ($Se^0$), which has low toxicity and excellent bioavailability (Torres et al. 2012). It's highly unstable, and it can readily revert to a dormant state. Encapsulation with chitosan, on the other hand, can help to stabilise it (Zhai et al. 2017).

The impact of dietary nano-selenium (nano-Se) with vitamin C on the growth of mahseer fish (*Tor putitora*) has recently been investigated (Khan et al. 2017). According to reports, nano-Se supplementation in feed up at the expense of 0.68 mg n-Se/kg dry feed greatly improved ($p < 0.05$) percent weight gain (percent WG), feed conversion efficiency (FCE percent), and specific growth rate (SGR) of fish when compared to fish fed a basal diet; however, the feed conversion ratio (FCR) was substantially lower ($p < 0.05$) in fish fed a supplemented diet when contrasted to fish fed a basal diet. In addition, a study on common carp (*Cyprinus carpio*) fed diets supplemented with nano-Selenium at 1 $mgkg^{-1}$ dry feed level reportedly significantly improved growth performance (in terms of final weight and weight gain) along with higher selenium contents in liver and muscle in fish fed nano-Selenium at 1 $mgkg^{-1}$ dry feed level compared to the control (Ashouri et al. 2015). Total protein and globulin levels were significantly greater ($p < 0.05$), but albumin levels were decreased for supplemented levels of 2 mg $kg^{-1}$ dry feed. In comparison to diets containing higher or lower amounts of nano-Se supplementation, antioxidant activities in fish given 1 and 2 $mgkg^{-1}$ dry feed were generally considerably ($p < 0.05$) improved.

### 8.3.2 Iron (Fe)

Iron has a vital part in the physiological processes of oxygen transport, cell respiration, lipid oxidation reactions, immune system function, and infection defence in fish (Beisel 1982; Hilty et al. 2011; Tahri et al. 2016). Because most natural bulk iron sources have limited solubility and bioavailability, dietary iron supplementation for fish is required to meet dietary needs (Hilty et al. 2011). The effects of iron oxide nanoparticles ($nFe_2O_3$) on growth in freshwater shrimp (*Macrobranchium rosenbergii*) post-larvae were investigated in comparison to a control diet. The fish fed supplemental doses of 10–20 mg $nFe_2O_3$ $Kg^{-1}$ dry feed had significantly better survival, growth, digestive enzyme activities, body biochemical composition, and some haematological parameters than those fed basal feed ($p < 0.05$) (Tahri et al. 2016). However, negative responses were recorded at greater supplemented amounts, such as 30–50 mg $nFe_2O_3$ $Kg^{-1}$ dry feed. Previously, researchers compared the performance of Indian major carp (*Labeo rohita*) on a base diet and two other diets (basal feed supplemented with 0.54 mg $nFe_2O_3$ $Kg^{-1}$feed and 0.55 mg

$FeSO.7H_2O$ $Kg^{-1}$feed, respectively) (Behera et al. 2014). They found that fish fed a nano-iron supplemented diet had significantly better survivability, growth, and haematological parameters than fish fed a control diet, but there was no significant difference in performance between fish fed diets containing 0.54 mg $nFe_2O_3$ $Kg^{-1}$feed and 0.55 mg $FeSO.7H_2O$ $Kg^{-1}$feed. When freshwater fish, *Oreochromis niloticus*, were exposed to biologically generated $Fe_2O_3$ NPs at concentrations of 0.5, 5, and 10 g/ml, some haematological parameters (RBC, WBC, Hb, and HCT) were reported to be considerably negatively affected (Chris et al. 2018). When an Indian major carp (Labeo rohita) was exposed to an environment (water) containing 500 mg of $nFe_2O_3$/litre of water for 25 days, blood parameters were impacted differentially. Throughout the duration, fish in the control had greater white blood cell (WBC) counts but lower red blood cell (RBC) counts than those in the treated water (Remya et al. 2015). The mean corpuscular haemoglobin concentration (MCHC) levels in both the treatment and control groups were relatively consistent throughout the trial duration. Other blood parameters, such as mean corpuscular volume (MCV), mean corpuscular haemoglobin (MCH), and haematocrit (Hct), were greater in the treatment for the first few days but then dropped in favour of the control. (Saravanan et al. 2015) obtained similar results after exposing Indian major carp (*Labeo rohita*) to different doses (1 and 25 mg/l) of $nFe_2O_3$ for 96 h. The control and the two treatments reacted differentially to various haematological parameters.

### 8.3.3 *Silver (Ag)*

Silver nanoparticles (AgNPs) are very well-known nanoparticles for their commercialization worldwide, owing to their unique biological properties, and are frequently used in medicine. Despite AgNPs' biological activities and a vast range of applications, there is a paucity of research on their effects on health and the environment. The widespread usage of AgNPs around the world, as well as their release into the aquatic ecosystem, has sparked fears of a negative influence on aquatic life. Silver nanoparticles have a critical role in the cleaning of water in which fish habitat systems are preserved due to the natural environment. Silver nanoparticles (AgNPs) are the most extensively studied multi-mechanism nano-based antibacterials. These nanoparticles contain specialised surfaces that attach to receptors in living cells, allowing them to cure and diagnose specific diseases. Silver ions ($Ag^+$) are released and bind to bacterial cell membrane proteins, causing cell membrane rupture and death. The antibacterial activity of chitosan-Ag nanocomposites (CAgNCs) against the fish disease *Aliivibrio salmonicida* has been examined. Previous research has shown that silver-based nanoparticles reduced *A. salmonicida* growth at 50 and 100 mg/L, respectively, indicating a minimum inhibitory concentration (MIC) and minimum bactericidal concentration (MBC) (Dananjaya et al. 2016).

Different nanoparticles have been created specifically for testing their action at the cellular level. When compared to other nanoparticles, silver nanoparticles are securely bound to the majority of bacteria, resulting in antibacterial activity. Furthermore, multiple studies have demonstrated that AgNPs are efficient against pathogenic organisms such as *B. subtilis*, *Vibrio cholerae*, and *Escherichia coli*, and that higher surface areas of AgNPs enable better contact with bacteria (Zinjarde 2012). Plants such as *Carica papaya* and *Musa paradisiacal* (banana) have been designed and tested to combat *Aeromonas hydrophila* (Munawar 2021). In juvenile *Feneropenaeus indicus*, biogenic Ag-NP made from tea leaf extract (*Camellia sinensis*) showed bactericidal efficacy against *Vibrio harveyi*, but only at high dosages of the nanoparticles (Onitsuka et al. 2019). Synthesised nanoparticles containing *Carica papaya* (papaya) have antibacterial action at a concentration of 153.6 g $mL^{-1}$. Biogenic CuO NPs were evaluated against *Aeromonas hydrophila, Pseudomonas fluorescens*, and *Flavobacterium branchiophilum* in 2015 and showed improved antibacterial activity against all fish pathogens even at lower doses, i.e. above 20 μ g/mL (Kumar et al. 2015).

### *8.3.4 Zinc (Zn)*

Zinc is required for the growth and health of all higher animals, including fish. Zinc helps animals grow, functions as an antibiotic, and regulates their immune and reproductive systems. Zn cannot be stored in the body of an animal, and a lack of it can cause repeated infections, a lack of appetite, and issues with taste and smell. Zinc supplementation is consequently required on a regular basis (Zalewski et al. 2005; Case and Carlson 2002). The effects of nano zinc oxide (nZnO) on grass carp (*Ctepharyngodon idella*) growth and haematological parameters have been studied in comparison to ZnO and $ZnSO_4$ as dietary zinc supplements in basal feed (Faiz et al. 2015). For each treatment, two amounts of supplementation (30 (level 1) and 60 (level 2) mg $Kg^{-1}$) were evaluated. Fish fed level 1 of nZnO supplemented diets had significantly higher percent weight gain (percent WG), specific growth rate (SGR), and feed conversion ratio (FCR) than fish fed level 2 of nZnO supplemented diets ($p < 0.05$). Growth was slowed in fish fed both levels of $ZnSO_4$ and level 2 of ZnO, according to the findings. Supplementing with $ZnSO_4$ and ZnO at both levels, as well as nZnO level 2, reduced haematological parameters including red blood cells (RBCs) and white blood cells (WBCs). This implies that the best dietary supplementation level of nZnO for *C. idella* is 30 mg/kg feed (Haghtalab et al. 2015; Akhtar et al. 2012).

## 8.4 Nanoparticle Applications in Fisheries and Aquaculture

Nanotechnology is already being used in the food business, with researchers interested in how NMs may affect food structure, texture, and quality, as well as technical applications in food production, processing, storage, transportation, and traceability (Chaudhry et al. 2008; Tiede et al. 2008). In the realm of fish and shellfish farming, there is a slew of possibilities. Fresh fish's perishability has long been a source of concern; therefore, any packaging that might extend its shelf life would be beneficial. This can be accomplished in a variety of ways. First, nanopolymers and coatings for packaging reinforcement are available, which may lower the occurrence of bruising or mechanical damage to packed fish fillets (De Azeredo 2009). Nanopackaging, unlike most of the conventional plastics, can be manufactured from natural nanoscale materials such as cellulose, starch, or chitosan and is thus expected to be biodegradable (De Azeredo 2009; Thompson et al. 2004). It has also been proposed that the meat sector use strong and light nanopackaging (Lee 2010). Microbial activity causes fresh fish items to spoil. Antimicrobial and antifungal surfaces can be used in packaging, and nanosilver is known for its antibacterial qualities (De Azeredo 2009; Moraru et al. 2003). Silver nanoparticles (AgNPs), zinc oxide (ZnO), as well as magnesium oxide (MgO), nano-silica (nS), aluminium oxide ($Al_2O_3$), titanium dioxide ($TiO_2$), and copper oxide (CuO) aid in a variety of bioactivities and biomodifications, and thus have a broad array of applications in agriculture, fisheries, food preservation and packaging, natural fibre strengthening, and the removal of contaminants from soil and water bodies, among others (Muthiah et al. 2019) (Fig. 8.1).

### *8.4.1 Water Treatment*

One of the most critical pillars for long-term aquaculture is water treatment. In recent years, water contamination has been identified as the world's most serious health threat, and it is steadily increasing as a result of waste dumping from cities, businesses, and agriculture, as well as the misuse of antibiotics and other synthetic substances in fisheries. The degradation of waters in this way has a direct impact on human health by reducing the availability of clean groundwater, but it also has an indirect impact on aquatic creatures, whose ingestion can result in a variety of food-borne illnesses. Apart from that, the fishery industry suffers a significant financial loss as a result of microorganisms and heavy metals in these waters, which induce growth retardation and fish death. Nanotechnology has important applications in aquaculture, such as water treatment to offer a favourable and safe environment for fish spawning. In this light, the scientific community supports adsorption and photocatalysis as the most efficient and cost-effective methods of water purification.

To eliminate pollutants from water, ultrathin nano-scale particles have also been utilised. Trichloroethane, carbon tetra chloride, and polychlorinated biphenyl, for

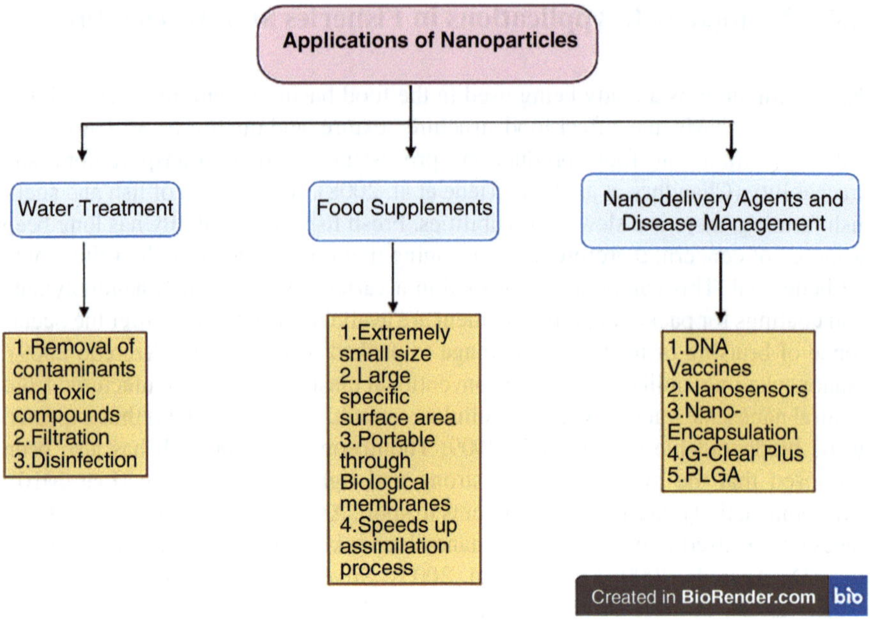


**Fig. 8.1** Different types of nanoparticles applications in the field of aquaculture. (Created with BioRender.com)

example, can be effectively removed from the environment using iron-derived ultrathin nanopowder (Wang et al. 2018). Researchers created and effectively implemented nanocrystalline zinc oxide film for photocatalytic water degradation and purification (Aal et al. 2009), as well as titanium oxide nanoparticles for coating ceramic or stone, which can remove moss and bacteria from fish bowls used for fish rearing (Van Doorslaer et al. 2012). Similarly, due to the improved stability of the nanoparticles on the foam, which aids in the controlled release of the material to the environment when desired, low-cost silver nanoparticles coated on polyurethane foam have been proposed for the effective killing of bacteria or providing an antibacterial filter for water treatment (Jain and Pradeep 2005). As a result, silver nanoparticles can be termed "new-generation antimicrobials" (Rai et al. 2009). The use of nanosilver-coated zeolite to reduce fungal development during commercial fish raising and the use of a 0.5% nanosilver filter to prevent fungal infections greatly reduced fungal infections (Johari et al. 2016). The different types of nanoparticles used in the current application of nanomaterials for water purification and filtration can be expressed. Multiwall nanotubes, single-wall nanotubes, metals and metal oxides, silica polymers, and composites are examples of nanoparticles (Aitken et al. 2006). Nano-check, a cleaning substance based on nanotechnology, has been successfully used to clean fish ponds and swimming pools. Nano-check is made up of nanoparticles (40 nm) of a lanthanum-based compound that absorbs phosphorus from water and prevents algae growth, reducing nutrient enrichment in water bodies (Rivas-Aravena et al. 2015). As a result, excessive algae growth in the pond can be controlled by limiting phosphorus availability, extending the pond's life. Furthermore, nanoscale-based weedicides have already been created and can aid in specifically targeting the tissues of the pest, as well as improved product delivery, which has been found to be very effective for large water bodies (Can et al. 2011). Iron oxide nanoparticles, according to (Karnik et al. 2005), can be used to improve ultra-filtration technology. The use of iron oxide nanoparticles for cleaning water for consumption has been proposed; additionally, the mode of action is primarily enhanced disinfection, which could be an effective strategy for most water supply providers. "Nano-863" is another novel material used in water management. Nano-863 has been used successfully for antibacterial effects, algal growth prevention, disease control, and pollution control (Shiwen et al. 2015). Furthermore, chitosan and chitosan-based nanoparticles were shown to have a number of positive effects on water management (Udo et al. 2018) (Table 8.1).

Magnetic konjac glucomannan (KGM) aerogels were produced by one of the study groups to cleanse water from arsenite (Ye et al. 2016). The pH-dependent capacity of the planned system was discovered, along with green step characteristics. However, graphene oxide and graphene nanosheets (GNs) have received a lot of interest in recent years because of their important function in eliminating various pollutants from water (Liu et al. 2016; Kuang et al. 2017; Motamedi et al. 2014). It has been particularly concentrated on the fabrication of hybrid GO-TiO2 for various environmental and energy applications, such as adsorption, and evacuating heavy metal ions and organic dyes from wastewater (Hu et al. 2013; Atchudan et al. 2017). Thanks to the research community for taking an interest in this topic of water treatment and contamination removal. Because, if left unattended, their higher concentrations can have negative, even life-threatening effects on human health by accumulating in the tissues of aquatic animals, particularly fish at the bottom of the aquatic food chain, whose consumption is highly recommended to combat cardiovascular diseases (CVDs) and cancer (Sioen et al. 2007). Fish are the most vulnerable and exposed aquatic creatures since they feed and live in an aquatic environment. They have no protection from the harmful effects of these contaminants (Saleh and Marie 2015; Mahboob et al. 2014). According to certain research, natural processes (e.g. volcanic activity) or anthropogenic acts cause increased levels of heavy metal (Hg, Cd, and Pb) buildup in the tissues of marine aquatic creatures (Dugo et al. 2006). Similarly, $F^{-1}$ toxicity was revealed to be responsible for the experimental fish's enzyme activities, gastrointestinal function, and immune system failure

**Table 8.1** Water pollutant removal methods based on nanoparticles

| System | Target | References |
|---|---|---|
| AgNP-coated polyurethane foam | *Escherichia coli* | Jain and Pradeep (2005) |
| IAO/GO | $F^{-1}$ | Liu et al. (2016) |
| 3-D RGO hydrogel | Hg and $F^{-1}$ | Wu et al. (2016) |
| NCC | $NO^{-3}$ | Azadbakht et al. (2016) |
| FeOOH-GO nanocomposites | $F^{-1}$ | Kuang et al. (2017) |
| $TiO_2$ and $TiO_2$–$SiO_2$ | $F^{-1}$ | Zheng et al. (2016) |

(Manna et al. 2007), as well as the freshwater snail *Physella acuta's* habitat degradation and destruction (Camargo and Alonso 2017). Wu et al. used a 3D RGO (three-dimensional reduced graphene oxide) hydrogel produced by a hydrothermal technique for $Hg^{+2}$ and $F^{-1}$ removal from aqueous solutions in this study (Wu et al. 2016). Their findings revealed that the aerogel has a significant capacity for the adsorption of $Hg^{+2}$ and $F^{-1}$, with 185 and 31.3 mg $g^{-1}$, respectively. They recommended the method as a good way to regulate pollutants in the environment. Azadbakht and colleagues created nanocrystalline cellulose (NCC) to remove $NO_3$ from aqueous solutions (Azadbakht et al. 2016). They found that bagasse-based NNC could be a useful strategy for removing nitrate from both water and wastewater reservoirs after obtaining a peak level removal of 25% at pH 6. Liu et al. developed a selective adsorbent based on magnetic iron–aluminium oxide/graphene oxide (IAO/GO) NPs for water purification from $F^{-1}$ (Liu et al. 2016). The absorbent was found to exhibit improved $F^{-1}$ selective adsorption performance, stability in an acid-base environment, and super para-magnetism properties. As a result, they speculated that IAO/GO-based adsorbents could be useful for $F^{-1}$ in natural water resources. For the elimination of $F^{-1}$ from the aqueous solution, another study group used $TiO_2$ and $TiO_2$–$SiO_2$ nanocomposite. The manufactured adsorbents had very high levels of $F^{-1}$ adsorption, with $TiO_2$ adsorption reaching 94.3 mg $g^{-1}$. Liu et al. tested several nanosystems and found that nano net was one of the best, ensuring a 100% increase in fish survival rate. They also discovered a significant reduction in water nitrite and nitrate levels, as well as better pH and water quality.

### 8.4.2 *Food Supplements*

Nanotechnology is a new technique for increasing aquaculture productivity by developing and applying innovative nanoparticles, nanocomposites, and other substances. Nanomaterials used as supplements in fish feed have a significant impact on fish growth performance. Because of their extremely small size and large specific surface area, they can easily pass through most biological membranes without causing distortion, assisting in effective absorption. Dietary minerals of a nanoscale size can move through cells faster than bulk particles, speeding up the assimilation process and improving not only the fish's growth but also their overall health (Onuegbu et al. 2018). Iron nanoparticle addition in fish diets, for example, increased the growth of sturgeon and carp by 30 and 24%, respectively. Similarly, adding nanoselenium and selenomethionine nanoparticles to a fish meal at a dose-dependent dosage increased *C. auratus'* growth, muscle development, and antioxidant status (Srinivasan et al. 2016). Further research has demonstrated that supplementing the diet with elemental nanoparticles such as selenium, iron, zinc, and silver can boost overall fish development performance (Can et al. 2011; Shiwen et al. 2015). Nano-863 is another new substance that has recently become popular in China for improving aquaculture. Nano-863 is made by combining nanoparticles with strong light-absorbing capabilities and sintering them with a ceramic carrier at high temperatures. It has a variety of uses in aquaculture, including increasing fish activity and energy, increasing shrimp appetite, and promoting aquatic creature growth and development (Shiwen et al. 2015). The emergence and application of nanotechnology in their manufacturing and production are being re-engineered to get maximum benefit in improving the physical, chemical, and nutritional quality of feed and their respective constituents (Navrotsky 2003). Nanoparticles made up of various compounds are now being researched as growth promoters. Nanoparticles will improve fish feed by boosting the proportion of fish meal ingredients that pass through the gut tissue, facilitating efficient digestion, rather than passing directly through the fish digestive system, according to Onuegbu et al. (2018). Many aquaculture professionals are looking into feed supplementation as a way to improve fish productivity by enhancing the physical, chemical, and nutritional quality of the feed. Fish production has shown a considerable increase in survivorship and growth performance when supplied with dietary nanoparticles (Onuegbu et al. 2018) like iron and zinc in the case of sturgeon and carp (Srinivasan et al. 2016) and copper (20 mg/kg) in the case of *M. rosenbergii* (Muralisankar et al. 2016) as compared to control feed administration. Selenium is another key fish feed additive for increased development and reproduction, with studies showing that selenium nanoparticles promote body weight gain and improve antioxidant properties in a variety of fish species. In comparison to the basal diet, nano-selenium supplementation up to 0.68 mg/kg of dry fish feed significantly improved % weight gain, feed conversion efficiency, and specific growth rate of fish ($p < 0.05$). In a similar study, nano-selenium supplementation at 1 mg/kg of dry fish feed resulted in enhanced growth and greater selenium levels in the liver and muscle of *C. carpio* compared to the control (Ashouri et al. 2015). Additional benefits of selenium nanoparticles in fish production have been demonstrated directly through improved growth performance and indirectly through stress resilience and health enhancement, such as improved growth performance in Crucian carp (*Carassius auratus gibelio*) (Nastova et al. 2014; Zhou et al. 2009); enhanced growth performance and reduced oxidative stress in juvenile Grass carp (*Ctenopharyngodon idellus*) (Johari et al. 2016); enhanced growth performance, muscle composure, and reduced oxidative stress in juvenile Grass carp (*Ctenopharyngodon idellus*) (Sarkar et al. 2015); enhanced growth performance, and glutathione peroxidase activity in Barramundi (*Lates calcarifer Block*) (Fotedar and Munilkumar 2016). In addition, selenium functions as a chemopreventive agent in fish (Rajendran 2013) and helps to maintain the integrity of the gill membrane (Halver 2007). Furthermore, the synergistic effects of selenium supplementation on blood enzyme activities, antioxidant response, and assimilation in *B. barbus* (Kouba et al. 2014), as well as the improved nutritional value of tilapia (*Oreochromis niloticus*) (Molnár et al. 2012) and the growth performance and yield of tilapia fortified with vitamins C and E (Fonseca et al. 2013), have been noted. In addition to promoting fish growth, selenium has also been shown to lower carp mortality. In comparison to the control group (17.5%), low mortality of carps in the ponds was seen (0.7%) when a selenium-enhanced diet was added, suggesting the importance of selenium in boosting survivorship (Nastova et al. 2014). According to recent studies, food supplementation with nano zinc and copper resulted in improved survival and growth in *M. rosenbergii (*Srinivasan et al. 2016*)*. Metallic nanoparticles, including iron, iron oxide, selenium, zinc, copper, silver, and magnesium oxide, are among the many additional nutrients that have been used in aquaculture development. The optimal level of such nanoparticles for feed supplementation can be proposed based on such investigations (Table 8.2).

Supplementing gilthead seabream (*Sparus aurata*) with selenium (Se), zinc (Zn), and manganese (Mn) NPs in their early weaning meals increased stress resistance and bone mineralization (Izquierdo et al. 2017). A formulation of solid lipid (SL) NP-encapsulated 6-COUM showed enhanced uptake of the compound by two gilthead seabream (*Sparus aurata L.*) cell types, namely an established cell line (SAF-1 cells) and primary cultures of head-kidney cells, when compared to a competitor's 6-coumarin loaded pectin microparticles (MPs) (HK). As a result, SLNPs can be used as nanocarriers to transport biologically active compounds to fish (Trapani et al. 2015). In rainbow trout, a meal supplemented with iron NPs and *Lactobacillus casei* as a probiotic considerably enhanced growth parameters (Mohammadi and Tukmechi 2015), whereas a diet supplemented with 16 mg $kg^{-1}$ MnO NPs significantly increased growth and the antioxidant defence system of freshwater prawns (*Macrobrachium rosenbergii*) (Asaikkutti et al. 2016). In the same way, copper (Cu) NP supplementation at 20 mg $kg^{-1}$ significantly increased the growth and biochemical constituents of freshwater prawn (*Macrobrachium rosenbergii*) post-larvae (Muralisankar et al. 2016) and red sea bream *(Pagrus major*) (El Basuini et al. 2017). In a freshwater common carp fish, Kunjiappan et al. (Kunjiappan et al. 2015) investigated the hepatoprotective and antioxidant properties of *Azolla microphylla*-based gold NPs (GNPs) against acetaminophen (APAP)-induced toxicity (*Cyprinus carpio L.*). GNPs improved metabolic enzymes, hepatotoxic indicators, oxidative stress markers, altered tissue enzymes, reduced hepatic ions, aberrant liver histology, and other factors. As a result, phytochemically produced GNaP from *Azolla microphylla* was indicated as an effective protector against acetaminophen-induced liver damage in freshwater common carp fish (Kunjiappan et al. 2015). Sharif Rohani and colleagues investigated the effects of three different amounts of aloe vera NPs (0.5%, 1%, and 1.5% of the diet) on Siberian sturgeon growth, survival, and body composition (*Acipenser baerii*). Their findings revealed that a diet supplemented with 1% aloe vera NPs considerably improved Siberian sturgeon growth indicators when compared to controls (Sharif Rohani et al. 2017). Another study looked at the effects of ginger (GN) and GNNPs on performance, cognition, immunity, and the prevention of motile *Aeromonas septicaemia* (MAS) in *Cyprinus carpio* fingerlings in the same year. Fish fed 1 and 0.5 g GNNPs per kg feed had 100% relative percentage survival (RPS), whereas fish fed 0.5 g GN per kg feed had a 20% mortality rate and a 71% RPS. These findings demonstrated that GNNPs, rather than GN, are a successful formulation for preventing MAS (Korni and Khalil 2017). AgNPs made from *Azadirachta indica* (neem) were created to test their immunomodulatory function in *Cirrhinus mrigala* fingerlings challenged with *Aeromonas hydrophila*. It was found that these NPs have potential immunomodulatory and antibacterial action after dramatically increasing the functional activity of immunological markers in fish treated with them (Rather et al. 2017). Erdem and his colleagues

**Table 8.2** Studies regarding supplement delivery using nanoparticles in aquatic animals

| Nanoparticle | Mode of administration | Target | Experimental design | Observations | References |
|---|---|---|---|---|---|
| **Se NP (+/-Vit E)** | Extruded into pellets | Juvenile Mahseer fish | Vitamin C doses of 100, 200, and 300 mg $kg^{-1}$ were combined with a nano se dosage of 0.68 mg $kg^{-1}$ to examine if they had any synergistic effects on growth, feeding, or physiological measures | Blood physicochemical properties such as Hb, WBC, RBC count enhanced, and liver and muscle protein content rose when vitamin C300 + nano se 0.68 mg kg1 was used | Khan et al. (2017) |
| **Se NP** | Extruded feed was added and air dried | Juvenile common carp | The growth performance of fish was evaluated at three different doses of SeNP (0.5, 1, and 2 mg/kg dry food) | Carp development was enhanced by 1 mg nano-se per kg feed, while 2 mg/kg caused toxicity | Ashouri et al. (2015) |
| **$Fe_3O_4$ NP** | Aloe vera leaf extract was used to make AgNP, which was combined with normal feed | Rainbow trout, *Labeo rohita* | Hematological parameters, respiratory burst activity, and serum bactericidal activity were assessed using 0.5 mg of $Fe_3O_4$ NP (T1) and $Fe_2SO_4$ (T2) per kg dry diet | Control had a lower Fe content in muscle than T1 and T2, whereas $Fe_3O_4$ NP had the greatest Fe level. In FeNP-treated diets, RBCs and haemoglobin levels, immunological parameters, bactericidal activity, and myeloperoxidase activity were all increased | Behera et al. (2014) |
| **CuNP** | The dough containing ZnNP was baked at 105 °C, chilled, and then pelletised using egg albumin and cod liver oil | *Macrobrachium rosenbergii* | Cu-NPs were added to the baseline meals at 0, 10, 20, 40, 60, and 80 mg $kg^{-1}$. *M. rosenbergii* PL was given these cu-NPs supplemented meals for 90 days | Survival, growth, digestive enzyme activities, biochemical constituent concentrations, and total and differential haemocyte count of prawns fed 20 mg (200 nm) cu-NPs $kg^{-1}$ supplemented feed exhibited significant ($P < 0.05$) improvements, whereas 40–80 mg cu-NPs $kg^{-1}$ supplemented feed showed negative performance | Muralisankar et al. (2016) |

(continued)

**Table 8.2** (continued)

| Nanoparticle | Mode of administration | Target | Experimental design | Observations | References |
|---|---|---|---|---|---|
| **SeNP** | In a drum mixer, se sources were slowly combined with the diet ingredients part by part, then extruded and air-dried at room temperature | *Carassius auratus gibelio* | T-1- (se NP), T-2, selenomethionine, and control were the three treatments. The relative growth rate as well as the final weight were calculated | In comparison to T-1 and T-2, the control had lower se content in muscle, while se NP had the greatest se content in muscle. Dietary interventions had no effect on survival rate or feed conversion ratio | Zhou et al. (2009) |
| **Se, Zn and Mn NPs** | Microdiets made by pelletizing, drying, grinding, and sieving squid powder with water-soluble components, vitamins, and gelatin | Seabream larvae | Larvae were fed one of the experimental diets (with no minerals, organic and inorganic form of the mineral, and nano forms of the mineral) every 45 minutes for 24 days after being fed rotifers supplemented with DHA protein. Image analysis was used to evaluate diet acceptability | Inorganic versions of these minerals were shown to be less efficient than organic forms.<br>In compared to a non-supplemented diet, minerals improve larval weight and bone mineralization. Furthermore, the larvae were less stress tolerant, and the fish had more bone abnormalities in the pre-hemal area. Nanometal additions of Zn, Mn, and se had little effect on growth.<br>However, stress resistance and bone mineralization have improved | Izquierdo et al. (2017) |
| **$Mn_3O_4$ NP** | Added to pelletised feed and dried at 40 °C | *Macrobrachium rosenbergii* | Seven groups of *M. rosenbergii* PL (1.42 ± 0.35 cm length; 0.18 ± 0.02 g weight) were studied with different levels of Mn NP for 90 days | The results of the trial showed that prawns fed a meal supplemented with 3–18 mg Mn-oxide NPs/kg had improved growth performance, final weight, and FCR ($P < 0.05$) | Asaikkutti et al. (2016) |

(continued)

recently synthesised AgNPs from *Aeromonas sobria* to test their antibacterial efficacy against a variety of fish infections (*H. alvei, P. rettgeri, M. morganii subsp. sibonii, C. braakii, E. hermannii, A. hydrophila, E. cloacae*, and *E. coli*). These NPs were thought to be a promising candidate for use as a disinfectant or antibacterial agent for better fish health management due to their great efficacy against *A. hydrophila* (Erdem et al. 2019).

Dietary iron oxide and copper nanoparticles (20 mg/kg) have been suggested for improving *M. rosenbergii* growth and health (Srinivasan et al. 2016; Muralisankar et al. 2016). When compared to the basal diet given to the African catfish, *Clarias gariepinus*, chitosan-derived nanoparticles (5 mg/kg) produced improved growth and feed utilisation (Udo et al. 2018). In addition to using dietary nanoparticles as fish feed supplements, there are several commercially patented goods that are employed as well. One such feed addition is "nano PUFA," which is intended to boost daily weight gain in fish and shrimp. Nano PUFA is made up of colloidal polyunsaturated fatty acid particles embedded in amino acid matrices and then bio-encapsulated for use in fish and shrimp nutrition. It also contains essential enzymes, extracts from a variety of helpful herbs, and feed-grade polymers (Moges et al. 2020).

### 8.4.3 Nanodelivery Agents and Disease Management

Nanomedicines include nanodiagnostics, nanovaccines, nanotherapeutics, and nanotoxicological compounds. Nanotechnologies enable new ways for strong fish health diagnostics, even down to the single virus level, as well as delivery mechanisms, particularly for heat-sensitive or labile medications. Various laboratories have researched the effects of silver and zinc oxide nanoparticles as promising antibacterial and antifungal agents against fish pathogens (bacteria, fungi, viruses, etc.) and diagnostic and therapeutic agents. Aquaculture is one of the most disease-prone systems, resulting in insecure development and long-term viability. It is also vulnerable to novel infections that can cause huge damage. As a result, focused vaccine/drug administration is critical for sustaining aquatic creature health. Several antibiotics are commonly employed as therapeutic medicines in aquaculture, either through feed or injection. The primary stumbling block is the drug's bioavailability, which makes it difficult for it to reach the fish underwater (Plant and Lapatra 2011). Because of the fish's low immune response, the traditional technique of drug/vaccine delivery to them has several drawbacks. Certain carriers, like oil/water emulsions, have been employed to improve the efficiency of medications, vaccinations, and nutrients; however, this strategy has resulted in numerous side effects. Nanoencapsulation is a revolutionary strategy for boosting aquaculture productivity by mitigating the negative impacts of oil emulsions. A vaccination or medicine is encapsulated inside another chemical that acts as a nanoscale protective shell in this approach. Nanoencapsulated complexes are extremely durable, able to withstand extreme temperatures and variations in water acidity, and hence are advantageous to use (Dominguez 2014).

Nanomaterial-based technologies could successfully prevent and manage illnesses and infections, boosting aquaculture productivity by a factor of ten. Nanomaterial-based technologies could help aquaculture grow by preventing and controlling illnesses and infections. Some of the fish health uses of nanotechnology include antibacterial or antifungal surfaces created utilising porous nanostructures, nanosensors in aquaculture systems for detecting infections in water, and the nanodelivery of veterinary products and fish pharmaceuticals through fish meals. Nano-trace elements have a far higher utilisation rate, up to 100%, than typical inorganic trace elements, which is normally extremely limited because the latter might enter the animal body by direct penetration. Epizootic Ulcerative Syndrome (EUS) (Kuan et al. 2013) and vibriosis in fish, as well as white spot syndrome (WSS) in shrimp species (Rajeshkumar et al. 2009), are some of the most frequent diseases caused by viruses, bacteria, and fungi. Because nanomaterials act at the same scale as viruses or disease-infecting particles, controlling these diseases would necessitate early detection and elimination of pathogens. Since nanoencapsulated materials are robust and can survive high temperatures or acidity, chitosan-based wrap-around vaccines as nanoencapsulation carriers for effective treatment delivery against bacteria and viruses that cause fish ailments have been described. Such nanomaterials can be used to produce pathogen-free fish seedlings as well as shrimp or prawn post-larvae for commercial farming.

Nanomedicine is a fast-expanding area of nanotechnology, and there is potential to employ these advancements to monitor and enhance fish health (Freitas Jr 2005). Due to the poor stability of pharmaceuticals in natural water, many fish treatments must be administered through food, or it must be accepted that much of any aqueous treatment will be washed away. NMs have been used to develop new medication delivery methods for people, and they could potentially be employed to develop veterinary medicines for fish. Solid core drug delivery systems (SCDDS) include coating a solid NP with a fatty acid shell to keep the medication of interest contained. This approach is particularly beneficial for heat-sensitive or labile medicines since it operates at low temperatures and pressures (Trivedi et al. 2014). Drug distribution can also be accomplished using porous NMs. Mesoporous silica particles, for example, can be utilised to release chemicals in a controlled manner (Strømme et al. 2009). This latter approach could be used to deliver fish vaccinations, for example. Pathogen detection nanosensors are also becoming available. Electrical nanosensors can already detect single virus particles, indicating that nanosensors will play a significant role in illness detection in the near future (Patolsky et al. 2004).

Because of their bactericidal properties, silver nanoparticles have been used in disease control (Gong et al. 2007). One of the most difficult difficulties in aquaculture was infection and disease control caused by viruses, bacteria, fungi, and parasites. Antimicrobials have traditionally been used in aquaculture to combat bacterial illnesses. Excessive use of these substances, on the other hand, resulted in resistant strains, making the therapies ineffective (Pelgrift and Friedman 2013). In a previous investigation of resistance strains in fish producers in 25 nations, it was discovered that tetracycline was the most commonly used antibiotic. In addition, several isolated bacteria from tilapia demonstrate a broad antibiotic range. *Aeromonas salmonicida*, *Photobacterium damselae*, *Yersinia ruckeri*, *Listeria* sp., *Vibrio* sp., *Pseudomonas* sp., and *Edwardsiella* sp. were among the resistant strains (Tuševljak et al. 2013). Some researchers researched the silver nanoparticle effect for various approaches to managing significant pathogens from fishes, mollusks, and crustaceans to explore other choices to avoid different ailments in aquatic organisms (Swain et al. 2014; Seil and Webster 2012). Silver nanoparticles were tested for antibacterial activity against two fish infections, *Lactococcus garvieae* and *Streptococcus iniae*, in a recent study (Raissy and Ansari 2011). It was determined what the minimum inhibitory concentration (MIC) and minimum bactericide concentration (BC) were. According to the findings, the MIC ranges from 1.12 to 5 g $mL^{-1}$ for *L. garvieae* and 1.2 to 2.5 g $mL^{-1}$ for isolated *S. iniae*. For *L. garvieae* and *S. iniae*, the mean MIC was 2.59 and 2.1, respectively. The strain *S. iniae* was shown to be more sensitive to silver nanoparticles than the strain *L. garvieae*. In another study (Saleh et al. 2017), the antiparasitic impact of AgNPs was assessed against *Ichthyophthirius multifiliis*, a parasite that causes white spot in freshwater fishes, and it was discovered that 10 and 5 $ngL^{-1}$ of AgNPS have antiparasitic effects in vitro and in vivo investigations.

The application of AgNPs synthetised from natural products to *Vibrio harveyi* control in infected *Feneropenaeus indicus* organisms yielded positive results (Vaseeharan et al. 2010), for example, when AgNPs were synthesised by *Camellia sinensis* to control *Vibrio harveyi* in infected *Feneropenaeus indic*us organisms. In vivo experiments revealed that a dose of 10 g $mL^{-1}$ suppressed bacterial growth in 70% of cases. In another work (Sivaramasamy et al. 2016), *Bacillus subtilis*, a non-pathogenic bacteria, was utilised to synthesise a nano molecule, which was then tested for antimicrobial activity against *V. parahaemolyticus* and *V. harveyi* in *Litopenaeus vannamei*-infected mice. The survival rate in the control group was 1%, but it was 90% in the nano compound group. It was also discovered that when bacteria and plant extracts were employed, no physiological changes or harmful effects on the shrimps were seen. It was observed that AgNPs encapsulated with starch and applied through immersion baths (20 min) with 10 ng of nanoparticle concentrations had antiparasitic and antifungal effects in *Carassius auratus* infected with *Ichthyophthirius multifiliis* and *Aphanomyces invadans*. The findings revealed that fish recovered after 3 days without becoming harmful as a result of the AgNPs treatment (Barakat et al. 2016). In vitro, some researchers tested AgNPs encapsulated with chitosan (75 g $mL^{-1}$) for antibacterial activity against *Vibrio tapetis* (Dananjaya et al. 2014). The lowest inhibitory concentration of one nano compound comprising chitosan and AgNPs was determined in 2017 to evaluate the antifungal efficacy in vitro against *Fusarium oxysporum* (Dananjaya et al. 2017). Later, in vivo studies were performed on fungus-infected zebrafish. In vitro experiments revealed that a concentration of 250 g $mL^{-1}$ of nano compound had a minimum inhibitory concentration of 38.69%, while in vivo tests revealed fungal mycelium degradation storage in fish backs, confirming that fungus was damaged by nano compound contact.

Nanoparticles encapsulating the medicine will improve the specificity of the target region while also assisting in better drug absorption. Researchers have recently had some success creating nanoencapsulated vaccinations against the bacteria *L.anguillarum* in Asian carp and the white spot syndrome virus in shrimp (Rajesh Kumar et al. 2008, 2009). Furthermore, particular medicines such as antibiotics, vaccines, probiotics, and nutraceuticals can be tagged on nanoparticles in a controlled time way. G-Clear Plus is a brand-new probiotic supplement designed to improve probiotic delivery in fish (Moges et al. 2020). It's a collection of probiotic strains that help fight the nasty gut sickness caused by several *Vibrio* bacteria. G-Clear Plus works in principle by selectively eliminating infections while simultaneously colonising the probiotic strains used in G-Clear Plus. G-Clear Plus is organic in nature because it does not contain antibiotics or powerful chemicals. Nanoparticles such as chitosan and poly-lactide glycolipid acid (PLGA) can also be employed to create nanovaccines (Dubey et al. 2016). Because of their mild inflammatory response in fish and shrimps, these chemicals can be employed to effectively remove bacteria as well as destroy viral infections. Nanocapsules containing nanoparticles can efficiently be utilised to retain bioavailability for a longer period of time in the event of mass vaccination, where vaccine delivery is a critical component. In addition, chitosan nanoparticles have been employed in the production of vaccines, such as the inactive virus vaccine against infectious salmon anaemia virus (ISAV), which uses the ISAV gene sequences as an adjuvant (Rivas-Aravena et al. 2015). This vaccine provided up to 77% protection against the ISAV virus. Rajesh Kumar et al. (2008) have shown that chitosan nanoparticles may be used to successfully deliver DNA vaccination to fish (Rajesh Kumar et al. 2008). In addition, using chitosan and chitosan/tripolyphosphate nanoparticles, the same group developed an oral DNA vaccine against *Vibrio anguillarum* in Asian sea bass (*Lates calcarifer*); the results showed that sea bass orally vaccinated with the chitosan-DNA (pVAOMP38) complex showed moderate (46%) protection against experimental *V. anguillarum* infection (Rajesh Kumar et al. 2008). Furthermore, an oral vaccine against the reddish body iridovirus, which infects the turbot *Scophthalmus maximus*, has been created and proven to be efficacious (Zheng et al. 2016). Feed supply is another challenge in aquaculture that necessitates efficient distribution. Because water is the medium through which fish reproduce, feed, and prosper, keeping the integrity and quality of feed in water is difficult due to its high liquidity. Because some nanoparticles can modify the physical qualities of feed, such as buoyancy and hardness, using nano-capsulation technology to preserve fat-soluble vitamins, minerals, and fatty acids in water may be helpful. The use of nanoencapsulation technology in the application of carotenoids, trace minerals, vitamins, and fatty acids has been proposed, with the primary purpose of enhancing bioavailability (Khot et al. 2012). Nutrients, medicines, food antimicrobials, and food additives can all be carried in them (Reza Mozafari et al. 2008). Nanoparticles have also been employed for enhanced and rapid diagnosis of fish disease, in addition to drug/vaccine development. Gold nanoparticles have played a crucial role in this regard. For the rapid diagnosis of furnaculosis infection in fish, gold nanoparticles coupled with *Aeromonas salmonicida* antibodies were utilised (Saleh et al. 2015). Early detection of yellow head virus in shrimp (Khunthong et al. 2013) and spring viremia of carp virus (Saleh et al. 2012) has also been achieved utilising specific colorimetric tests based on gold nanoparticles. Nanoparticles are also utilised in delivery methods, such as anaesthesia and oil emulsifiers (Kheawfu et al. 2017). When compared to current conventional approaches to blending, such advanced techniques of nano-based delivery systems have several advantages, including a broader application for nutraceuticals, less chance of leakage into the water, a conservative amount, protection against oxidation, superior efficiency of antioxidants and preservatives, high bioavailability of compounds, and thermal and pH stability (Aklakur et al. 2016).

## 8.5 Challenges and Limitations

Despite its many uses, nanotechnology's usage in aquaculture is currently limited and has not been fully commercialised to the extent that it has been in other industrial sectors. Academics bear the brunt of nanotechnology research breakthroughs, while industries are limited to patent ownership. Furthermore, the size of scientific patents has grown significantly, necessitating the inclusion of broad claims when filing in order to ensure future freedom to operate in the face of changing economic conditions. Another disadvantage of nanotech goods is the high initial capital investment required due to the usage of advanced equipment and services, which can be overcome if the job is done on a large scale. Start-ups and small businesses may become demotivated as a result of this. There are, however, specific financing bodies that can help start-ups establish their setting by providing a seed fund to demonstrate the principle of scale-up. Furthermore, most metal-based nanoparticles are nonbiodegradable and have the ability to increase biomagnification. Several metal-based nanoparticles, including silver, gold, titanium, and magnesium, have been shown to have dose-dependent harmful effects on animal development and to cause developmental defects in young animals (Hegde et al. 2016; Iswarya et al. 2019; McGillicuddy et al. 2017; Verma et al. 2018). Metal nanoparticles can also interact with organisms' essential enzymes, causing negative biological effects. In this context, zinc oxide nanoparticles were shown to have specific clinical manifestations such as a pulmonary inflammatory response in mice at high doses (>1 mg/mL) (Sayes et al. 2007), as well as severe symptoms of lethargy, anorexia, vomiting, diarrhoea, body weight loss, and even death when gastrointestinally administered, whereas such effects were mitigated in bulk-sized zinc particles at equal doses (Wang et al. 2006). As a result, increased use of such nanoparticles may eventually be hazardous to living organisms. In this context, the use of organic or carbon-based nanoparticles appears promising and has the potential to mitigate the harmful effects of metal nanoparticles, thereby reducing environmental impact. Carbon-dots, graphene nanoparticles, metal nanoparticles encapsulated in graphene/silica, and carbon-metal-nano-conjugates are examples of possible nanotechnology advancements that could help enhance agriculture while minimising environmental effects. Despite these drawbacks, the introduction of nanotech goods to the market will yield significant benefits due to their cost effectiveness, specificity, stability, mobility, and variety of applications. Risk assessment using in vitro and in vivo models,

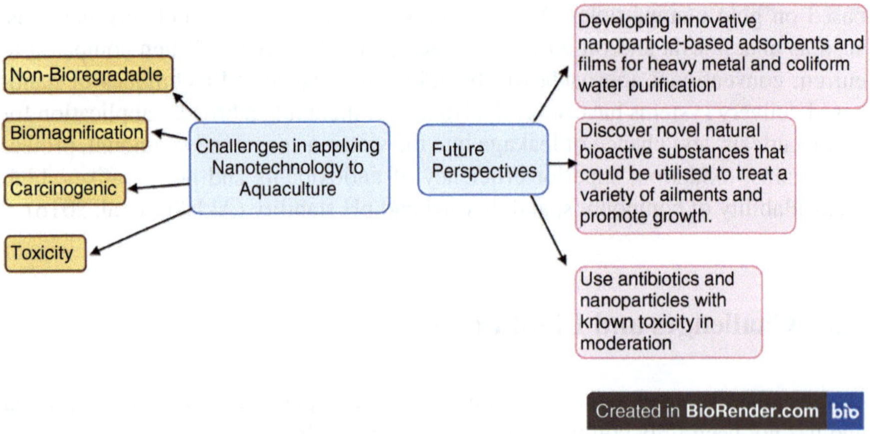


**Fig. 8.2** Challenges and limitations in the application of nanoparticles in aquaculture and its future perspectives. (Created with BioRender.com)

appropriate dose determination for field application with minimal environmental impact, shifting to organic and environmentally friendly nanoparticles, and evaluating the use of nanoparticles in controlled clinical trials using appropriate models are all possible approaches for mitigating deleterious effects and selecting environmentally friendly nanoparticles. As a result, advances in nanoscience will have far-reaching implications for aquaculture; nevertheless, producing next-generation fish with nanotechnology will be difficult, necessitating additional research into biological behaviour, specificity, scalability, and practical application (Fig. 8.2).

## 8.6 Conclusion

Nanotechnology undoubtedly plays an important role in the growth and long-term viability of aquaculture. Various nanotechnology-based solutions have been used to enhance the major pillars of aquaculture and fishing up to this point. Despite their enormous potential, the use of watershed management ideas and methods, as well as nanotechnology in fisheries and aquaculture, is still limited. In comparison to consumer goods and human healthcare, nanotechnological uses in fisheries, agriculture, and water management as a whole are significantly less prevalent. Only a small number of nano-products or nanotechnologies are now accessible in the fisheries subsector, despite their enormous potential. This position opens up a lot of possibilities for allocating more money to nanotechnology research and implementation in the subsector.

However, there is rising worry in this field concerning the toxicity of NPs, as well as the overuse of antibiotics and other synthetic substances. As a result, the use of safe and environmentally friendly methods is unavoidable. Based on this knowledge, we highly suggest that future research in aquaculture and fisheries be focused on:

(i) Developing innovative nanoparticle-based adsorbents and films for heavy metal and coliform water purification.
(ii) Using antibiotics and nanoparticles with known toxicity in moderation.
(iii) Discovering novel natural bioactive substances that could be utilised to treat a variety of ailments and promote growth.

## References


Aal AA, Mahmoud SA, Aboul-Gheit AK (2009) Nanocrystalline ZnO thin film for photocatalytic purification of water. Mater Sci Eng C 29(3):831–835

Aitken RJ et al (2006) Manufacture and use of nanomaterials: current status in the UK and global trends. Occup Med 56(5):300–306

Akhtar MJ et al (2012) Zinc oxide nanoparticles selectively induce apoptosis in human cancer cells through reactive oxygen species. Int J Nanomedicine 7:845

Aklakur M, Asharf Rather M, Kumar N (2016) Nanodelivery: an emerging avenue for nutraceuticals and drug delivery. Crit Rev Food Sci Nutr 56(14):2352–2361

Ali TE-S et al (2014) Structural and functional effects of early exposure to 4-nonylphenol on gonadal development of Nile tilapia (Oreochromis niloticus): a-histological alterations in ovaries. Fish Physiol Biochem 40(5):1509–1519

Ali TE-S et al (2018) Effects of dietary Biogen and sodium butyrate on hematological parameters, immune response, and histological characteristics of Nile tilapia (Oreochromis niloticus) fingerlings. Aquac Int 26(1):139–150

Asaikkutti A et al (2016) Dietary supplementation of green synthesized manganese-oxide nanoparticles and its effect on growth performance, muscle composition and digestive enzyme activities of the giant freshwater prawn Macrobrachium rosenbergii. J Trace Elem Med Biol 35:7–17

Ashouri S et al (2015) Effects of different levels of dietary selenium nanoparticles on growth performance, muscle composition, blood biochemical profiles and antioxidant status of common carp (Cyprinus carpio). Aquaculture 446:25–29

Atchudan R et al (2017) Effective photocatalytic degradation of anthropogenic dyes using graphene oxide grafting titanium dioxide nanoparticles under UV-light irradiation. J Photochem Photobiol A Chem 333:92–104

Azadbakht P et al (2016) Removal of nitrate from aqueous solution using nanocrystalline cellulose. Int J Environ Health Eng 5(1):17

Bae WK et al (2009) Highly efficient green-light-emitting diodes based on CdSe@ ZnS quantum dots with a chemical-composition gradient. Adv Mater 21(17):1690–1694

Barakat KM, El-Sayed HS, Gohar YM (2016) Protective effect of squilla chitosan–silver nanoparticles for Dicentrarchus labrax larvae infected with vibrio anguillarum. Int Aquac Res 8(2):179–189

Behera T et al (2014) Nano-Fe as feed additive improves the hematological and immunological parameters of fish, Labeo rohita H. Appl Nanosci 4(6):687–694

Beisel WR (1982) Single nutrients and immunity

Boxall AB, et al (2007) Current and future predicted environmental exposure to engineered nanoparticles. Central Science Laboratory, Department of the Environment and Rural Affairs, London, p 89

Camargo JA, Alonso Á (2017) Ecotoxicological assessment of the impact of fluoride (F−) and turbidity on the freshwater snail Physella acuta in a polluted river receiving an industrial effluent. Environ Sci Pollut Res 24(18):15667–15677

Can E et al (2011) Nanotechnological applications in aquaculture-seafood industries and adverse effects of nanoparticles on environment. J Mater Sci Eng 5(5)

Case C, Carlson M (2002) Effect of feeding organic and inorganic sources of additional zinc on growth performance and zinc balance in nursery pigs. J Anim Sci 80(7):1917–1924
Chaudhry Q et al (2008) Applications and implications of nanotechnologies for the food sector. Food Addit Contam 25(3):241–258
Chris UO, Singh N, Agarwal A (2018) Nanoparticles as feed supplement on growth behaviour of cultured catfish (Clarias gariepinus) fingerlings. Mater Today: Proc 5(3):9076–9081
Citarasu T (2010) Herbal biomedicines: a new opportunity for aquaculture industry. Aquac Int 18(3):403–414
Dananjaya S et al (2014) Chitosan silver nano composites (CAgNCs) as potential antibacterial agent to control Vibrio tapetis. J Vet Sci Technol 5(5)
Dananjaya S et al (2016) Antimicrobial effects of chitosan silver nano composites (CAgNCs) on fish pathogenic Aliivibrio (vibrio) salmonicida. Aquaculture 450:422–430
Dananjaya S et al (2017) Comparative study on antifungal activities of chitosan nanoparticles and chitosan silver nano composites against fusarium oxysporum species complex. Int J Biol Macromol 105:478–488
Dawood MA et al (2020) Marine-derived chitosan nanoparticles improved the intestinal histo-morphometrical features in association with the health and immune response of Grey Mullet (Liza ramada). Mar Drugs 18(12):611
De Azeredo HM (2009) Nanocomposites for food packaging applications. Food Res Int 42(9):1240–1253
Dominguez GA (2014) Nanotechnology in the aquaculture industry. Sustainable Nano
Dubey S et al (2016) Aeromonas hydrophila OmpW PLGA nanoparticle Oral vaccine shows a dose-dependent protective immunity in Rohu (Labeo rohita). Vaccines (Basel) 4(2)
Dugo G et al (2006) Concentration of Cd (II), Cu (II), Pb (II), Se (IV) and Zn (II) in cultured sea bass (Dicentrarchus labrax) tissues from Tyrrhenian Sea and Sicilian Sea by derivative stripping potentiometry. Food Control 17(2):146–152
El Basuini MF et al (2017) Effects of dietary copper nanoparticles and vitamin C supplementations on growth performance, immune response and stress resistance of red sea bream, Pagrus major. Aquac Nutr 23(6):1329–1340
El-Hammady A, El-Kasheif M, Ibrahim S (2007) Synergistic reactions between vitamin eand selenium in diets of hybrid tilapia (oreochromis nilot1cus x oreochromis aureus) and their effect on the growth and liver histological structure. Egypt J Aquat Biol Fish 11(1):53–81
Erdem B et al (2019) Biosynthesis of silver nanoparticles from Aeromonas sobria and antibacterial activity against fish pathogens. Int J Environ Sci Technol 16(9):5125–5130
Faiz H et al (2015) Zinc oxide, zinc sulfate and zinc oxide nanoparticles as source of dietary zinc: comparative effects on growth and hematological indices of juvenile grass carp (Ctenopharyngodon idella). Int J Agric Biol 17(3):568
FAO (2018) The State of World Fisheries and Aquaculture 2018. FAO, p. 227
Fonseca SBD et al (2013) Influence of levels and forms of selenium associated with levels of vitamins C and E on the performance, yield and composition of tilapia fillet. Food Sci Technol 33:109–115
Fotedar R, Munilkumar S (2016) Effects of organic selenium supplementation on growth, glutathione peroxidase activity and histopathology in juvenile barramundi (Lates calcarifer Bloch 1970) fed high lupin meal-based diets. Aquaculture 457:15–23
Freitas RA Jr (2005) What is nanomedicine? Nanomedicine: nanotechnology. Biol Med 1(1):2–9
George S et al (2012) Surface defects on plate-shaped silver nanoparticles contribute to its hazard potential in a fish gill cell line and zebrafish embryos. ACS Nano 6(5):3745–3759
Gong P et al (2007) Preparation and antibacterial activity of Fe3O4@ Ag nanoparticles. Nanotechnology 18(28):285604
Haghtalab A, Mohammadi M, Fakhroueian Z (2015) Absorption and solubility measurement of $CO_2$ in water-based ZnO and $SiO_2$ nanofluids. Fluid Phase Equilib 392:33–42

Halver J (2007) Selenium in fish nutrition: a role in gill membrane integrity. in Nutritional biotechnology in the feed and food industries. In: Proceedings of Alltech's 23rd Annual Symposium. The new energy crisis: food, feed or fuel? Alltech

Handy RD et al (2008a) Manufactured nanoparticles: their uptake and effects on fish – a mechanistic analysis. Ecotoxicology 17(5):396–409

Handy RD et al (2008b) The ecotoxicology and chemistry of manufactured nanoparticles. Ecotoxicology 17(4):287–314

Hegde K et al (2016) Current understandings of toxicity, risks and regulations of engineered nanoparticles with respect to environmental microorganisms. Nanotechnol Environ Eng 1(1):5

Hillyer JF, Albrecht RM (2001) Gastrointestinal persorption and tissue distribution of differently sized colloidal gold nanoparticles. J Pharm Sci 90(12):1927–1936

Hilty FM et al (2011) Incorporation of Mg and Ca into nanostructured Fe2O3 improves Fe solubility in dilute acid and sensory characteristics in foods. J Food Sci 76(1):N2–N10

Hosseini SF et al (2016) Development of bioactive fish gelatin/chitosan nanoparticles composite films with antimicrobial properties. Food Chem 194:1266–1274

Hu C et al (2013) A brief review of graphene–metal oxide composites synthesis and applications in photocatalysis. J Chinese Adv Mater Soc 1(1):21–39

Iravani S, Zolfaghari B (2013) Green synthesis of silver nanoparticles using Pinus eldarica bark extract. Biomed Res Int 2013:1

Iswarya V et al (2019) Toxic effect of different types of titanium dioxide nanoparticles on Ceriodaphnia dubia in a freshwater system. Environ Sci Pollut Res 26(12):11998–12013

Izquierdo MS et al (2017) Organic, inorganic and nanoparticles of Se, Zn and Mn in early weaning diets for gilthead seabream (Sparus aurata; Linnaeus, 1758). Aquac Res 48(6):2852–2867

Jain P, Pradeep T (2005) Potential of silver nanoparticle-coated polyurethane foam as an antibacterial water filter. Biotechnol Bioeng 90(1):59–63

Johari SA et al (2016) Application of nanosilver-coated zeolite as water filter media for fungal disinfection of rainbow trout (Oncorhynchus mykiss) eggs. Aquac Int 24(1):23–38

Ju-Nam Y, Lead JR (2008) Manufactured nanoparticles: an overview of their chemistry, interactions and potential environmental implications. Sci Total Environ 400(1–3):396–414

Karnik BS et al (2005) Fabrication of catalytic membranes for the treatment of drinking water using combined ozonation and ultrafiltration. Environ Sci Technol 39(19):7656–7661

Khan MS et al (2015) Toxicity of silver nanoparticles in fish: a critical review. J Bio Environ Sci 6(5):211–227

Khan KU et al (2017) Synergistic effects of dietary nano selenium and vitamin C on growth, feeding, and physiological parameters of mahseer fish (tor putitora). Aquacult Rep 5:70–75

Kheawfu K et al (2017) Nanoemulsion: a suitable nanodelivery system of clove oil for anesthetizing Nile tilapia. Drug Discover Ther 11(4):181–185

Khosravi-Katuli K et al (2017) Effects of nanoparticles in species of aquaculture interest. Environ Sci Pollut Res 24(21):17326–17346

Khot LR et al (2012) Applications of nanomaterials in agricultural production and crop protection: a review. Crop Prot 35:64–70

Khunthong S et al (2013) Rapid and sensitive detection of shrimp yellow head virus by loop-mediated isothermal amplification combined with a lateral flow dipstick. J Virol Methods 188(1–2):51–56

Khurana A et al (2019) Therapeutic applications of selenium nanoparticles. Biomed Pharmacother 111:802–812

Kok B et al (2020) Fish as feed: using economic allocation to quantify the fish in: fish out ratio of major fed aquaculture species. Aquaculture 528:735474

Korni F, Khalil F (2017) Effect of ginger and its nanoparticles on growth performance, cognition capability, immunity and prevention of motile Aeromonas septicaemia in Cyprinus carpio fingerlings. Aquac Nutr 23(6):1492–1499

Kouba A et al (2014) Supplementation with sodium selenite and selenium-enriched microalgae biomass show varying effects on blood enzymes activities, antioxidant response, and accumulation in common barbel (Barbus barbus). Biomed Res Int 2014:1

Kuan GC et al (2013) Gold-nanoparticle based electrochemical DNA sensor for the detection of fish pathogen Aphanomyces invadans. Talanta 117:312–317

Kuang L et al (2017) FeOOH-graphene oxide nanocomposites for fluoride removal from water: acetate mediated nano FeOOH growth and adsorption mechanism. J Colloid Interface Sci 490:259–269

Kumar PV et al (2015) Green synthesis of copper oxide nanoparticles using Aloe vera leaf extract and its antibacterial activity against fish bacterial pathogens. BioNanoScience 5(3):135–139

Kumar N, Krishnani KK, Singh NP (2018) Effect of dietary zinc-nanoparticles on growth performance, anti-oxidative and immunological status of fish reared under multiple stressors. Biol Trace Elem Res 186(1):267–278

Kunjiappan S, Bhattacharjee C, Chowdhury R (2015) Hepatoprotective and antioxidant effects of Azolla microphylla based gold nanoparticles against acetaminophen induced toxicity in a fresh water common carp fish (Cyprinus carpio L.)

Le K, Fotedar R, Partridge G (2014) Selenium and vitamin E interaction in the nutrition of yellowtail kingfish (S eriola lalandi): physiological and immune responses. Aquac Nutr 20(3):303–313

Lee KT (2010) Quality and safety aspects of meat products as affected by various physical manipulations of packaging materials. Meat Sci 86(1):138–150

Liu L et al (2016) One-step synthesis of magnetic iron–aluminum oxide/graphene oxide nanoparticles as a selective adsorbent for fluoride removal from aqueous solution. RSC Adv 6(13):10783–10791

Luis AIS et al (2019) Trends in aquaculture sciences: from now to use of nanotechnology for disease control. Rev Aquac 11(1):119–132

Mahboob S et al (2014) Tissue metal distribution and risk assessment for important fish species from Saudi Arabia. Bull Environ Contam Toxicol 92(1):61–66

Manna P, Sinha M, Sil PC (2007) A 43 kD protein isolated from the herb Cajanus indicus L attenuates sodium fluoride-induced hepatic and renal disorders in vivo. BMB Rep 40(3):382–395

McGillicuddy E et al (2017) Silver nanoparticles in the environment: sources, detection and ecotoxicology. Sci Total Environ 575:231–246

Moges FD et al (2020) Mechanistic insights into diverse nano-based strategies for aquaculture enhancement: a holistic review. Aquaculture 519:734770

Mohammadi N, Tukmechi A (2015) The effects of iron nanoparticles in combination with lactobacillus casei on growth parameters and probiotic counts in rainbow trout (Oncorhynchus mykiss) intestine. J Vet Res 70(1)

Molnár T et al (2012) Improving the nutritional value of Nile tilapia fillet by dietary selenium supplementation

Moraru CI et al (2003) Nanotechnology: a new frontier in food science understanding the special properties of materials of nanometer size will allow food scientists to design new, healthier, tastier, and safer foods. Nanotechnology 57(12):24–29

Motamedi E, Atouei MT, Kassaee M (2014) Comparison of nitrate removal from water via graphene oxide coated Fe, Ni and Co nanoparticles. Mater Res Bull 54:34–40

Munawar N (2021) Interaction and applications of nanoparticles in fishes and aquaculture. Sch Bull 7(6):150–155

Muralisankar T et al (2016) The effect of copper nanoparticles supplementation on freshwater prawn Macrobrachium rosenbergii post larvae. J Trace Elem Med Biol 34:39–49

Muthiah M, Chipps S, Ojasvi PR (2019) On the advanced technologies to enhance fisheries production and management. Acta Sci Agric 3:216–222

Nasr-Eldahan S et al (2021) A review article on nanotechnology in aquaculture sustainability as a novel tool in fish disease control. Aquac Int:1–22

Nastova R, Gjorgovska N, Kostov V (2014) Selenium supplementation in fish nutrition. Agro Life Sci J 3

Navrotsky A (2003) Materials and nanotechnology. J Frankl Inst 340(3):263–268

Onitsuka S, Hamada T, Okamura H (2019) Preparation of antimicrobial gold and silver nanoparticles from tea leaf extracts. Colloids Surf B: Biointerfaces 173:242–248

Onuegbu C, Aggarwal A, Singh N (2018), ZnO nanoparticles as feed supplement on growth performance of cultured African catfish fingerlings

Pandey RK, Prajapati VK (2018) Molecular and immunological toxic effects of nanoparticles. Int J Biol Macromol 107:1278–1293

Patolsky F et al (2004) Electrical detection of single viruses. Proc Natl Acad Sci 101(39):14017–14022

Pelgrift RY, Friedman AJ (2013) Nanotechnology as a therapeutic tool to combat microbial resistance. Adv Drug Deliv Rev 65(13–14):1803–1815

Plant KP, Lapatra SE (2011) Advances in fish vaccine delivery. Dev Comp Immunol 35(12):1256–1262

Prashanth L et al (2015) A review on role of essential trace elements in health and disease. J Dr NTR Univ Health Sci 4(2):75

Rai M, Yadav A, Gade A (2009) Silver nanoparticles as a new generation of antimicrobials. Biotechnol Adv 27(1):76–83

Raissy M, Ansari M (2011) In vitro antimicrobial effect of silver nanoparticles on Lactococcus garvieae and streptococcus iniae. Afr J Microbiol Res 5(25):4442–4445

Rajendran D (2013) Application of nano minerals in animal production system. Res J Biotechnol 8(3):1–3

Rajesh Kumar S et al (2008) Potential use of chitosan nanoparticles for oral delivery of DNA vaccine in Asian sea bass (Lates calcarifer) to protect from vibrio (Listonella) anguillarum. Fish Shellfish Immunol 25(1–2):47–56

Rajeshkumar S et al (2009) Oral delivery of DNA construct using chitosan nanoparticles to protect the shrimp from white spot syndrome virus (WSSV). Fish Shellfish Immunol 26(3):429–437

Rather MA et al (2017) Synthesis and characterization of Azadirachta indica constructed silver nanoparticles and their immunomodulatory activity in fish. Aquac Res 48(7):3742–3754

Rawn DF et al (2009) Analysis of fin and non-fin fish products for azamethiphos and dichlorvos residues from the Canadian retail market. Int J Food Sci Technol 44(8):1510–1516

Remya AS et al (2015) Iron oxide nanoparticles to an Indian major carp, Labeo rohita: impacts on hematology, iono regulation and gill Na+/K+ ATPase activity. J King Saud Univ Sci 27(2):151–160

Reza Mozafari M et al (2008) Nanoliposomes and their applications in food nanotechnology. J Liposome Res 18(4):309–327

Rivas-Aravena A et al (2015) Development of a nanoparticle-based oral vaccine for Atlantic salmon against ISAV using an alphavirus replicon as adjuvant. Fish Shellfish Immunol 45(1):157–166

Saleh YS, Marie M-AS (2015) Assessment of metal contamination in water, sediment, and tissues of Arius thalassinus fish from the Red Sea coast of Yemen and the potential human risk assessment. Environ Sci Pollut Res 22(7):5481–5490

Saleh M et al (2012) Direct detection of unamplified spring viraemia of carp virus RNA using unmodified gold nanoparticles. Dis Aquat Org 100(1):3–10

Saleh M, Soliman H, El-Matbouli M (2015) Gold nanoparticles as a potential tool for diagnosis of fish diseases. In: Veterinary infection biology: molecular diagnostics and high-throughput strategies. Springer, pp 245–252

Saleh M et al (2017) Antiprotozoal effects of metal nanoparticles against Ichthyophthirius multifiliis. Parasitology 144(13):1802–1810

Saravanan M et al (2015) Iron oxide nanoparticles induced alterations in haematological, biochemical and ionoregulatory responses of an Indian major carp Labeo rohita. J Nanopart Res 17(6):1–12

Sarkar B et al (2015) Selenium nanoparticles for stress-resilient fish and livestock. Nanoscale Res Lett 10(1):1–14

Sayes CM et al (2007) Comparative pulmonary toxicity assessments of C60 water suspensions in rats: few differences in fullerene toxicity in vivo in contrast to in vitro profiles. Nano Lett 7(8):2399–2406
Seil JT, Webster TJ (2012) Antimicrobial applications of nanotechnology: methods and literature. Int J Nanomedicine 7:2767
Shah BR, Mraz J (2020) Advances in nanotechnology for sustainable aquaculture and fisheries. Rev Aquac 12(2):925–942
Sharif Rohani M, Haghighi M, Bazari Moghaddam S (2017) Study on nanoparticles of Aloe vera extract on growth performance, survival rate and body composition in Siberian sturgeon (Acipenser baerii). Iran J Fish Sci 16(2):457–468
Shaw BJ, Handy RD (2011) Physiological effects of nanoparticles on fish: a comparison of nano-metals versus metal ions. Environ Int 37(6):1083–1097
Shi L et al (2011) Effect of sodium selenite, Se-yeast and nano-elemental selenium on growth performance, Se concentration and antioxidant status in growing male goats. Small Rumin Res 96(1):49–52
Shiwen H et al (2015) Nanotechnology in agriculture, livestock, and aquaculture in China. A review. Agron Sustain Dev 35:369–400
Sioen I et al (2007) Development of a nutrient database and distributions for use in a probabilistic risk–benefit analysis of human seafood consumption. J Food Compos Anal 20(8):662–670
Sivaramasamy E et al (2016) Enhancement of vibriosis resistance in Litopenaeus vannamei by supplementation of biomastered silver nanoparticles by Bacillus subtilis. J Nanomed Nanotechnol 7(352):2
Srinivasan V et al (2016) Effects of dietary iron oxide nanoparticles on the growth performance, biochemical constituents and physiological stress responses of the giant freshwater prawn Macrobrachium rosenbergii post-larvae. Int J Fis Aquat Stud 4(2):170–182
Stone V et al (2010) Nanomaterials for environmental studies: classification, reference material issues, and strategies for physico-chemical characterisation. Sci Total Environ 408(7):1745–1754
Strømme M et al (2009) Mesoporous silica-based nanomaterials for drug delivery: evaluation of structural properties associated with release rate. Wiley Interdiscip Rev Nanomed Nanobiotechnol 1(1):140–148
Swain P et al (2014) Antimicrobial activity of metal based nanoparticles against microbes associated with diseases in aquaculture. World J Microbiol Biotechnol 30(9):2491–2502
Tahri M et al (2016) Health status of the swim bladder of the European eel Anguilla anguilla in northeastern Algeria's Lake Oubeïra. Int J Fish Aquat Stud 4(1):364–369
Thompson RC et al (2004) Lost at sea: where is all the plastic? Science(Washington) 304(5672):838
Tiede K et al (2008) Detection and characterization of engineered nanoparticles in food and the environment. Food Addit Contam 25(7):795–821
Torres S et al (2012) Biosynthesis of selenium nanoparticles by Pantoea agglomerans and their antioxidant activity. J Nanopart Res 14(11):1–9
Trapani A et al (2015) In vitro characterization of 6-Coumarin loaded solid lipid nanoparticles and their uptake by immunocompetent fish cells. Colloids Surf B: Biointerfaces 127:79–88
Trivedi V, Bhomia R, Mitchell J (2014) The preparation of solid core drug delivery systems (SCDDS) for the biomolecular delivery
Tuševljak N et al (2013) Antimicrobial use and resistance in aquaculture: findings of a globally administered survey of aquaculture-allied professionals. Zoonoses Public Health 60(6):426–436
Udo IU, Etukudo U, Anwana U-IU (2018) Effects of chitosan and chitosan nanoparticles on water quality, growth performance, survival rate and meat quality of the African catfish, Clarias gariepinus. Nano 1(1):12–25
Van Doorslaer X et al (2012) TiO2 mediated heterogeneous photocatalytic degradation of moxifloxacin: operational variables and scavenger study. Appl Catal B Environ 111:150–156

Vaseeharan B, Ramasamy P, Chen J (2010) Antibacterial activity of silver nanoparticles (AgNps) synthesized by tea leaf extracts against pathogenic Vibrio harveyi and its protective efficacy on juvenile Feneropenaeus indicus. Lett Appl Microbiol 50(4):352–356

Verma SK et al (2018) Rapid novel facile biosynthesized silver nanoparticles from bacterial release induce biogenicity and concentration dependent in vivo cytotoxicity with embryonic zebrafish – a mechanistic insight. Toxicol Sci 161(1):125–138

Vimal S et al (2012) Synthesis and characterization of CS/TPP nanoparticles for oral delivery of gene in fish. Aquaculture 358:14–22

Wang B et al (2006) Acute toxicity of nano-and micro-scale zinc powder in healthy adult mice. Toxicol Lett 161(2):115–123

Wang Z et al (2018) Recent advances in nanoporous membranes for water purification. Nano 8(2):65

Wu S, Kong L, Liu J (2016) Removal of mercury and fluoride from aqueous solutions by three-dimensional reduced-graphene oxide aerogel. Res Chem Intermed 42(5):4513–4530

Ye S et al (2016) Development of Mag-FMBO in clay-reinforced KGM aerogels for arsenite removal. Int J Biol Macromol 87:77–84

Yue Y et al (2017) Interaction of silver nanoparticles with algae and fish cells: a side by side comparison. J Nanobiotechnol 15(1):1–11

Zalewski PD et al (2005) Zinc metabolism in airway epithelium and airway inflammation: basic mechanisms and clinical targets. A review. Pharmacol Ther 105(2):127–149

Zhai X et al (2017) Antioxidant capacities of the selenium nanoparticles stabilized by chitosan. J Nanobiotechnol 15(1):1–12

Zheng F et al (2016) Development of oral DNA vaccine based on chitosan nanoparticles for the immunization against reddish body iridovirus in turbots (Scophthalmus maximus). Aquaculture 452:263–271

Zhou X et al (2009) Effects of different dietary selenium sources (selenium nanoparticle and selenomethionine) on growth performance, muscle composition and glutathione peroxidase enzyme activity of crucian carp (Carassius auratus gibelio). Aquaculture 291(1–2):78–81

Zinjarde SS (2012) Bio-inspired nanomaterials and their applications as antimicrobial agents. Chronicles Young Sci 3(1):74

# Chapter 9
# Nanotechnologies in the Health Management of Aquatic Animal Diseases

**Nithianantham Sundar Raj, Vishnu Kirthi Arivarasan, Azeez Sait Sahul Hameed, and Thangaraj Raja Swaminathan**

## 9.1 Introduction

Aquaculture, also known as "underwater agriculture," is the fastest-growing food-producing sector in the world. With the growth of aquaculture, consumers from low- to high-income nations have benefited from year-round availability and access to aquatic foods, which are rich in protein and micronutrients (Belton and Thilsted 2014; Béné et al. 2016; Thilsted et al. 2016; Belton et al. 2020). Among the continents, Asia contributes more than 90% of the world's aquaculture production (Bondad-Reantaso et al. 2005). Aquaculture encompasses a very wide range of different aquatic farming practices, including those involving crustaceans, fish, seaweeds, molluscs, and other aquatic species. More than 14.5 million people are dependent on aquatic farming practices for their livelihoods, food security, especially in rural communities, and poverty alleviation such as income generation and employment (Kumar et al. 2015).

With the rapid development of aquaculture in India, disease outbreaks remain the biggest challenge to the sustainability of aquaculture production. Viral and bacterial pathogens are a chronic risk for the aquaculture sector and its highly intensive fish farming practices (Kennedy et al. 2016). On the other side, increased trade and

N. S. Raj · T. R. Swaminathan (✉)
Peninsular and Marine Fish Genetic Resources Centre, ICAR- National Bureau of Fish Genetic Resources, CMFRI Campus, Kochi, Kerala, India

V. K. Arivarasan
Department of Microbiology, School of Bioengineering and Biosciences, Lovely Professional University, Phagwara, Punjab, India

A. S. S. Hameed
Aquatic Animal Health Laboratory, C. Abdul Hakeem College, Melvisharam, Vellore, Tamil Nadu, India

 
V. K. Arivarasan et al. (eds.), *Nanotechnological Approaches to the Advancement of Innovations in Aquaculture*, Nanotechnology in the Life Sciences, https://doi.org/10.1007/978-3-031-15519-2_9

supply of seeds including fish fry and shrimp larvae are often of low quality and affected by diseases (Stentiford et al. 2012). Many aquaculture systems, however, still lack the motivation to meet sustainability criteria because their targeted markets do not reward producers, leading to poor management and causing severe mortalities and economic losses to the industry (Walker and Winton 2010). Outbreaks prompted by diseases increase the use of antibiotics and other chemicals, increasing environmental risks and the lack of adequate biosecurity control.

In India, several transboundary aquatic animal diseases have caused massive economic losses, which include the spread and outbreaks of new emerging viral diseases like cyprinid herpesvirus 2 (CyHV-2), carp edema virus (CEV), viral nervous necrosis (VNN), red sea bream iridovirus disease (RSIVD), infectious spleen and kidney necrosis virus (ISKNV), and tilapia lake virus (TiLV). A few viral diseases have also been reported in shrimps: monodon baculovirus (MBV), white spot syndrome virus (WSSV), infectious hypodermal and hematopoietic necrosis virus (IHHNV), hepatopancreatic parvo-like virus (HPV), white tail disease (WTD), and infectious myonecrosis virus (IMNV).

The term nanotechnology refers to a multidisciplinary field which can be applied to various aspects of applications, which include large-scale industries, medicine, agriculture, etc. The discovery of nanoparticles in this modern world has brought fields of research and applications towards each other, leading to an amalgam of sciences. The nanoparticles are the particles which are formed in the size range of 1–100 nm, and this changes the physical and chemical properties of their bulk counterparts (Abou El-Nour et al. 2010). In the application of nanotechnology to aquaculture, the role of nanotechnology diversifies into many forms. The effects of nanotechnology and nanoparticles on aquaculture can be found in drug delivery development, food packaging, and many more (Abou El-Nour et al. 2010; Sannino 2021).

## 9.2 Nanotechnology in Aquaculture

Nanomaterials and their technology can be utilized for many applications, including diagnostics and drug delivery. Some of the reports have been elaborated here; Saleh and El-Matbouli (2015) reported the gold nanoparticles to be used to develop and evaluate a specific and sensitive hybridization assay for direct and rapid detection of the highly infectious pathogen termed cyprinid herpesvirus 3. A duplex PCR was developed for the detection of the cyprinid herpesvirus 2 (Luo et al. 2014), a droplet digital PCR for the detection of the cyprinid herpesvirus 2 (Hao et al. 2016), and similarly, the loop-mediated isothermal amplification assay for the rapid detection of the cyprinid herpesvirus 2 in Gibel carp was also developed (Zhang et al. 2014).

There are reports of nanoparticles being utilized as drug delivery agents, which have proven to be very much successful. The DNA vaccines against sea bass nodavirus-based chitosan nanoparticles have shown good results (Vimal et al. 2014), and some suggested that PLGA-based nanoparticles could be used for drug delivery (Adomako et al. 2012; Embregts and Forlenza 2016; Kole et al. 2019b). The delivery systems and their framework improve solubility and bioavailability and also lead to sustained release of hydrophobic drugs, thereby protecting the chemical payload from degradation and facilitating the cellular and tissue targeting (Hu et al. 2018; Oroojalian et al. 2020; Venkatesan et al. 2013). The utilization of chitosan and polycaprolactone polymers for the delivery of ascorbic acid was reported (Luis et al. 2021). Many researchers have reported on the utilization of the vaccine delivery for fish (Abbas 2021; Adams 2019; Bedekar and Kole 2022; Heyerdahl et al. 2018; Huang et al. 2021; Jeong et al. 2020; Kayansamruaj et al. 2020; Somamoto and Nakanishi 2020; Thirumalaikumar et al. 2021; Wang et al. 2020; Zhang et al. 2020; Zhang et al. 2021; Zhao et al. 2020; Zhu et al. 2020). There are also reports on the nano-based delivery of drugs and chemicals for aquaculture applications (Ji et al. 2015; Kole et al. 2019a; Li et al. 2013; Thirumalaikumar et al. 2021; Thwaite et al. 2018; Zhang et al. 2020; Zhang et al. 2021; Zhu et al. 2020). There is growing interest in developing a more sustainable and targeted delivery system for drugs and vaccines for aquatic animals. The role of the nanoparticles can be utilized for the definite release of the drug payload in a sustained manner.

## 9.3 Fish Viral Diseases

### *9.3.1 Cyprinid Herpesvirus 2 (CyHV-2)*

Cyprinid herpesvirus 2 (CyHV-2) is an emerging pathogen of the alloherpesviridae and is also known as herpesviral haematopoietic necrosis virus (HVHNV) of goldfish or goldfish haematopoietic necrosis virus (GFHNV) (Hanson et al. 2011). CyHV-2 infection has been reported globally, including in Asia (Waltzek et al. 2009), Europe (Boitard et al. 2016; Doszpoly et al. 2011; Jeffery et al. 2007), North America (Goodwin et al. 2006), and Oceania (Becker et al. 2014). At present, CyHV-2 has been reported in 16 countries, including India (Sahoo et al. 2016). CyHV-2 belongs to the family Alloherpesviridae and the genus Cyprinivirus. CyHV-2 has an icosahedral capsid with a linear dsDNA of about 290 kbp and encodes approximately 150 genes (Davison et al. 2013). Host-range studies for CyHV-2 have been limited; previously, CyHV2 was considered a pathogen of goldfish. To date, it has been able to infect a much wider range of cyprinid species, including crucian carp (*Carassius carassius*) and Prussian carp (*C. gibelio*) (Xu et al. 2013; Fichi et al. 2016; Zhao et al. 2019).

CyHV-2 causes acute disease in all ages of goldfish, and mortality occurs while infection can reach almost 100% at temperatures between 15°C and 25°C (Goodwin et al. 2009; Davison et al. 2013). The entry of the virus is unknown; an infected fish can transmit the virus horizontally, and vertical transmission is also possible from parent fish to offspring (Goodwin et al. 2009). Diseased fish show no remarkable external signs except pale gills, which result from severe destruction of the hematopoietic tissues (Jung and Miyazaki 1995). The most severe gross changes in the infected fish are swollen kidneys, an empty intestine, and splenomegaly with white nodular lesions in the spleen (Jung and Miyazaki 1995; Jeffery et al. 2007). According to histological identification, infected cells may have enlarged nuclei with margination of chromatin and the presence of intranuclear inclusion bodies (Jeffery et al. 2007). Viral DNA may be detected for longer periods in healthy goldfish brood stock due to the latency of CyHV-2 (Goodwin et al. 2009). There is no commercial vaccine or effective treatment available for CyHV-2 (Thangaraj et al. 2021). The virus remains a serious threat to the Indian ornamental fish trade and food fish aquaculture. To reduce and control the spread of the virus, it is necessary to follow strict biosecurity measures and regulate fish trading.

### *9.3.2 Carp Edema Virus (CEV)*

Carp edema virus (CEV), which causes carp edema virus disease (CEVD) or koi sleepy disease (KSD), is an emerging disease of global concern. CEV is considered a potential risk for the koi trade and for global carp aquaculture, because of its wide distribution and potential virulence. CEV was first detected in koi carp in Japan in 1974 (Murakami 1976). Due to the global trade in live ornamental fish, the virus has spread to other countries such as the USA, Czech Republic, Austria, India, Germany, the Netherlands, Italy, China, Brazil, Poland, Hungary, and South Korea (Rehman et al. 2020). Koi carp culture in India is growing rapidly in many states, including Kerala, Tamil Nadu, West Bengal, and Maharashtra. Unfortunately, the outbreak of CEV has occurred in koi farms in Kerala, India (Swaminathan et al. 2016). Koi carp and common carp are known to be species susceptible to CEV, with high mortality rates during the outbreaks reaching up to 75–100% (Miyazaki et al. 2005; Way and Stone 2013).

Currently, limited information is known about the virus that causes KSD. Clinical signs of KSD are typical sleepy behaviour, enophthalmia, a generalized oedematous condition, and gill necrosis (Lewisch et al. 2015). CEV is an unclassified double-stranded DNA virus and belongs to family Poxviridae. The size of the immature and mature virion is about 416–450 nm and 300–400 × 250–400 nm in diameter, respectively (Adamek et al. 2017).

The clinical signs of KSD and KHVD are very similar, and the disease can be misidentified easily. KSD diagnosis relies on PCR, and real-time PCR assays have been developed and validated for the detection of CEV infection (Oyamatsu et al. 1997; Matras et al. 2019). Attempts have also been made to replicate the CEV *in vitro*, but employing the fish cell lines presently available has not been successful (Jung-Schroers et al. 2016; Lewisch et al. 2015; Swaminathan et al. 2016). Although CEV has been present in aquaculture for many years, there is still a lack of knowledge about its transmission among species, which is important to prevent the spread of CEV/KSD (Matras et al. 2019). Until now, there have been no treatments or prophylactic measures available for KSD. Extreme precautionary measures and regulations are necessary in the live fish trade in order to prevent the further spread of CEV infection (Swaminathan et al. 2016).

### 9.3.3 Viral Encephalopathy and Retinopathy (VER)

Viral encephalopathy and retinopathy or viral nervous necrosis is an OIE significant disease. Viral nervous necrosis (VNN) infects fishes with so-called piscine nodaviruses and belongs to the family Nodaviridae. First isolated Nodaviridae is the striped jack nervous necrosis virus (SJNNV) and genus *Betanodavirus* (Mori et al. 1992). The virus has been recorded in over 120 different fish species, and the mortality associated with the disease is severe, reaching 100% depending on age, with younger fish being more susceptible (Munday et al. 2002; Costa and Thompson 2016). The disease, designated viral nervous necrosis (VNN) when it was first described in 1990 (Yoshikoshi and Inoue 1990), is also known as viral encephalopathy and retinopathy (OIE 2003). Subsequently, the disease has been reported globally, including in India in 2005 (Azad et al. 2005).

The VNN is classified under the family Nodaviridae and the genus *Betanodavirus*. The virus consists of a larger genomic segment of single-stranded positive-sense RNA (composed of two segments), which encodes the RNA-dependent RNA polymerase. The coat protein is encoded by RNA 2 (1.4 kbp), and RNA 3 of 0.4 kb encodes the protein B2. It has non-enveloped isometric symmetry (icosahedral) and is about 30 nm in diameter. Till now, nodaviruses have been classified by different clades: the striped jack clade (SJNNV), the red-spotted *grouper* clade (RGNNV), the tiger puffer clade (TPNNV), the barfin flounder clade (BFNNV), *Dicentrarchus labrax* encephalitis virus (DLEV), the Japanese flounder nervous necrosis virus (JFNNV), *Lates calcarifer* encephalitis virus (LcEV), Atlantic halibut nodavirus (AHNV), and the Malabar grouper nervous necrosis virus (MGNNV). Based on the optimal growth temperatures, some genotypes have been classified, e.g., 25–30°C for the RGNNV genotype, 25–30°C for the SJNNV genotype, 20°C for the TPNNV genotype, and 15°C for the BFNNV genotype, and differ among the other genotypic variants.

Investigations on the possible routes of VNN infection by horizontal and vertical transmission are also possible. Horizontal transmission is the common mode of infection for betanodavirus, and it has been demonstrated in European sea bass (Peducasse et al. 1999; Skliris and Richards 1999) and by the cohabitation of sea bream (*Sparus aurata*), an asymptomatic carrier of fish nodavirus in sea bass (Castric et al. 2001). Vertical transmission of the virus has been identified in various fish species (Arimoto et al. 1992; Grotmol et al. 1999; Grotmol and Totland 2000; Breuil et al. 2002). The virus localizes in the brain, spinal cord, and retina of affected fish and is characterized by neurological abnormalities (erratic swimming, spiral movements with belly-up) and a distinct vacuolization of the nerve tissue (brain and retina) (OIE 2006). The distribution of VNN in surviving fish was not only in the CNS but also in other tissues like gonad, intestine, stomach, kidney and liver of brood stock carriers (Arimoto et al. 1992; Munday et al. 2002). Moreover, the virus could spread between farms through contaminated personnel and farm equipment; hence, adequate biosecurity measures should be ensured to prevent its spread. Recently, scientists (ICAR-CIBA) found the remedy for this disease in India, and they developed a recombinant vaccine for VNN affecting several fish species under the commercial name Nodavac-R.

### 9.3.4 Red Sea Bream Iridovirus (RSIV)

Red sea bream iridovirus (RSIV), in the genus *Megalocytivirus* (Kurita and Nakajima 2012), is a virus causing extensive economic loss to the aquaculture industry. RSIVD has caused severe mortality in many cultured and wild fish species of several countries (OIE 2019) including India. The first outbreak of this pathogen was reported in farmed Asian sea bass (*Lates calcarifer*) in 2018 (Girisha et al. 2020). RSIV infects more than 30 other species of cultured marine fish (Matsuoka et al. 1996; Kawakami and Nakajima 2002).

RSIV belongs to the family Iridoviridae and the genus Megalocytivirus. It has identified three genotypes: red sea bream iridovirus (RSIV), infectious spleen and kidney necrosis virus (ISKNV), and *turbot reddish* body iridovirus (TRBIV) (Subramaniam et al. 2012). RSIV is an enveloped virus with icosahedral virions that are found within the cytoplasm of enlarged cells, and the virion size is about 200–240 nm in diameter. It is a double-stranded DNA virus with a genome size of 111.15 KB (Shiu et al. 2018; Puneeth et al. 2021). RSIV may spread not only in the summer, when this disease is prevalent, but also in the winter. The horizontal mode of transmission is the route for causing RSIV infection, and vertical transmission of this virus infection is not known. Clinical signs of the infected fish include lethargy, helpless swimming, severe anemia, petechiae of the gills, and enlargement of the spleen, with 20–60% mortality. Histopathology is characterized by the development of enlarged cells in the spleen, heart, kidney, liver, and gills and the observation of basophilic characteristics in Giemsa staining.

An effective formalin-inactivated vaccine was developed by Nakajima et al. (1999). Currently, for this virus, the injectable vaccine is now commercially available in Japan for red sea bream, striped jack (*Pseudocaranx dentex*), Malabar grouper (*Epinephelus malabaricus*), and orange-spotted grouper (*Epinephelus coioides*). In a recent report from India, RSIV infection has been found in farmed fish (Girisha et al. 2020). There is a potential threat from this virus that can hamper the industry. Hence, active research must be carried out on RSIV for its prevention and spread in India.

### 9.3.5 Infectious Spleen and Kidney Necrosis Virus (ISKNV)

Infectious spleen and kidney necrosis virus (ISKNV) is an emerging and trans-boundary fish pathogen that causes large-scale mortalities in fish. ISKNV belongs to the family Iridoviridae and the genus Megalocytivirus and is closely related to RSIV. Among these, ISKNV causes both symptomatic and asymptomatic infections in about 50 different freshwater, brackish, and marine species, particularly during the summer (Kurita and Nakajima 2012).

Since the first report of ISKNV in Chinese mandarin fish, Siniperca chuatsi (He et al. 2002), ISKNV has been reported in Australia (Go and Whittington 2006), Korea (Song et al. 2008), Singapore (Jeong et al. 2008), Malaysia (Subramaniam et al. 2012), Germany (Jung-Schroers et al. 2016), and Indonesia (Sukenda et al. 2020). Recently, ISKNV was reported in asymptomatic exotic ornamental fishes in India (Girisha et al. 2021). Global live ornamental fish trade without the compliance of appropriate pathogen screening and quarantine serves as the major reason for the transboundary spread of ISKNV (Jung-Schroers et al. 2016; Rimmer et al. 2015).

The genome size of ISKNV is 111 kb, and it has about 124 open reading frames (ORFs) (He et al. 2001). The virus is icosahedral in shape, encapsulated with a nucleocapsid structure, and the size of the virus ranges between 110 and 150 nm (Shi et al. 2004; Jung-Schroers et al. 2016). Clinical signs of ISKNV-infected fish show symptoms such as being lethargic and exhibiting severe anemia, petechiae of the gills, enlargement of the spleen, abnormal swimming, ulceration, haemorrhages, and a darkened body. ISKNV can cause severe infection at various stages of the life cycle, and moreover, it can cause up to a 75% mortality rate. In particular, in India, ISKNV is currently considered an exotic pathogen. However, considering the ornamental trade, it is very important to have a monitoring and surveillance program to screen for emerging and re-emerging diseases and develop preventive and control measures.

## 9.4 Tilapia Lake Virus Disease

Recently, tilapia lake virus (TiLV) has been noted as an important infectious agent that may pose a serious threat to the global tilapia industry (Jansen et al. 2019; Mugimba et al. 2018; Pulido et al. 2019). TiLV has been taxonomically classified as a Tilapia tilapinevirus species, in the *Tilapinevirus* genus, Amnoonviridae family (ICTV 2018). To date, TiLV has been reported in 16 tilapia-farming countries, including Asia, Africa, the Middle East, and South and Central America (Taengphu et al. 2020). The disease is caused by a novel orthomyxo-like virus also known as TiLV (Bacharach et al. 2016), and mortality rates ranging from 80% to 90%, especially in fingerlings and juveniles, have been reported following infection with TiLV (Fathi et al. 2017; Surachetpong et al. 2017). The clinical signs of the disease are lethargy, haemorrhages, abdominal swelling, exophthalmia, and severe mortalities in farmed and wild populations of tilapia, which are characterized by syncytia formation in the liver of affected fish. The condition is known as syncytial hepatitis of tilapia (Ferguson et al. 2014).

TiLV virions are enveloped in a round or oval shape, 55–100 nm in diameter, and a single-stranded RNA virus which contains 10 segments encoding 10 proteins. The total viral genome size is about 10,323 kb (Eyngor et al. 2014; Bacharach et al. 2016; Del-Pozo et al. 2017). The genome organization and ultrastructural morphology of TiLV resemble those of other orthomyxoviruses (Del-Pozo et al. 2017; Eyngor et al. 2014). A total of 20 complete TiLV genome sequences from six different countries are available, including 10 from Thailand, 3 from Bangladesh, 2 each from the USA and Israel, and one each from Peru, Ecuador, and India (Verma et al. 2022). In experimental trials, TiLV infection has been reported in tilapia to be associated with 70–90% mortality by intra-peritoneal (i.p.) injection (Mugimba et al. 2019) and cohabitation (Liamnimitr et al. 2018). To date, TiLV affects several species of cultured tilapia cichlids, including Nile tilapia (Tattiyapong et al. 2017), red tilapia (Tattiyapong et al. 2017; Mugimba et al. 2019), Mozambique tilapia (Nanthini et al. 2019) hybrid tilapia (Mugimba et al. 2019; Amal et al. 2018), and several wild tilapia (Eyngor et al. 2014). A recent experimental study has reported that other than tilapia species, giant gourami (Jaemwimol et al. 2018) and wild tinfoil barb (Abdullah et al. 2018) have also been found to be susceptible to TiLV infection.

TiLV can be transmitted both horizontally and vertically. Horizontal transmission is considered to be the major route for the spread of TiLV, with the virus being transmitted to healthy fish through direct contact with the skin and mucus of infected fish or through cannibalism of moribund or dead fish (Liamnimitr et al. 2018; Tang et al. 2021). Vertical transmission of TiLV has also been observed in natural cases of infection with TiLV and following experimental infection (Dong et al. 2020).

However, TiLV has been reported as a cause of mass mortality in tilapia in India (Behera et al. 2018). Since it is a viral disease, therefore, there is no effective treatment for it. In response to this, active surveillance for the presence or absence of the virus/disease was carried out. Some precautionary measures include banning the importation of tilapia from TiLV-confirmed countries. Other effective measures for control of the disease include a combination of biosecurity measures, the breeding of fish with improved resistance, and the development of new vaccination against the TiLV infection disease.

## 9.5 Shrimp Viral Diseases

### *9.5.1 Monodon Baculovirus (MBV)*

Monodon baculovirus (MBV), the causative agent of spherical baculovirosis, was the first virus reported in penaeid shrimp, described by Lightner and Redman (1981) from *Penaeus monodon* in Taiwan. MBV should be reclassified as *P. monodon* nudivirus (PmNV) and reassigned to the Nudiviridae (Wang and Jehle 2009; Yang et al. 2014). MBV infects *P. monodon*, *P. merguiensis*, *P. semisulcatus*, *P. kerathurus*, *P. vannamei*, *P. esculentus*, *P. penicillatus*, *P. plebejus*, and *M. ensis*. MBV was found to infect up to 20% of populations of wild *P. monodon* in Asia (Manivannan et al. 2004; Leobert et al. 2008). MBV has been detected in *P. monodon* in Taiwan, the Philippines, Malaysia, French Polynesia, Hawaii, Kenya, Mexico, Singapore, Indonesia, Israel, and Thailand. In India, the prevalence of MBV and mortalities in hatcheries and farms associated with MBV have been reported (Ramasamy et al. 1995; Karunasagar and Karunasagar 1998). Artemia does not act as the mechanical carrier for MBV (Sarathi et al. 2008). Vijayan et al. (1995) reported the presence of MBV in *P. indicus* along the southeast coast of India. MBV can infect *Macrobrachium rosenbergii* and produce lesions in the hepatopancreas similar to those in *P. monodon* at the early stage of infection (Gangnonngiw et al. 2010). There are about nine distinctly different isolates obtained from shrimp samples, which confirmed the existence of variation in the strains of MBV in India reported by Suganthi et al. (2012).

Symptoms of the disease include a reduction in feeding and growth, reduction in activity, and dark outgrowth on the gill surface of the shrimp. Mortalities occur primarily among post-larvae in the hatchery, although disease may also occur in juvenile and adult prawns (Johnson and Lightner 1988). Adult brood stock can carry the virus and transmit MBV to offspring via horizontal transmission (Paynter et al. 1992), direct from the water column, or cannibalism, and it is believed, but not proven, that transmission can also be vertical from brood stock to offspring. Infection can result in substantial economic loss due to poor performance in growth and reduced survival of post-larvae—up to 90% at high densities. Furthermore, stress and overcrowding are the predisposing factors that may increase the severity of MBV infection (Lightner et al. 1983a).

The MBV particles are rod-shaped and replicate in the nucleus. These appear either free or within proteinaceous polyhedral occlusion bodies and contain DNA. The nucleocapsids of MBV measure 42±3 nm by 246±15 nm, while the enveloped virions are larger, measuring 75±4 nm by 324± 33 nm (Lightner et al. 1983a; Chen et al. 1989; Brock and Lightner 1990). However, little is known about the mechanism of MBV production in the host cell. Diagnosis of MBV depends upon the demonstration of MBV occlusion bodies in hypertrophied nuclei of anterior midgut epithelium and hepatopancreatic cells by direct light microscopy or standard H&E staining (Lightner and Redman 1998). However, the detection of occlusion bodies requires a high level of infection, thus these traditional methods have limited sensitivity. Applications of molecular techniques, including in situ hybridization (Poulos et al. 1994) and polymerase chain reaction (Belcher and Young 1998), have been developed as more sensitive methods for detecting MBV.

## 9.6 White Spot Syndrome Virus (WSSV)

The WSSV has been one of the major threats to the shrimp industry over the past two decades, which is responsible for huge economic loss in the shrimp culture industry not only in India but also worldwide. The global loss to the shrimp culture industry due to this virus has been estimated to be about US$ 10 billion (Stentiford et al. 2009). The loss in India alone has been estimated to be about several million dollars per year, and the loss continues to threaten the long-term sustenance of the shrimp industry in India. WSSV was first reported in 1992 in Taiwan, from where it spread to all shrimp-growing countries (Flegel 1997). Disease outbreaks can reach a cumulative mortality of up to 100% within 3–7 days of infection (Escobedo-Bonilla et al. 2008). This is a very fast-reproducing, widely spreading, and highly virulent crustacean pathogen. WSSV has a large host range, infecting shrimp, crayfish, and lobster, among many other species (Sánchez-Paz 2010). A total of 98 potential host species for WSSV have been identified from the scientific literature (Stentiford et al. 2009).

The virus belongs to the family Nimaviridae, the genus *Whispovirus*, and contains a double-stranded DNA genome with sizes ranging from 292.9 to 307.2 kb (Sánchez-Martínez et al. 2007). The intact virus is enveloped and elliptical in shape, measuring 266 nm. The nucleocapsid of WSSV is cylindrical in shape (420 nm), with one end flat and the other pointed, and has a pattern of opaque and transparent striations arranged perpendicularly to the long axis of the nucleocapsid (Sahul Hameed et al. 1998). The viral DNA encodes major structural proteins and minor proteins for various functions, including pathogenesis (Van Hulten et al. 2001a; Van Hulten et al. 2001b; Van Hulten et al. 2002).

Clinical signs of WSSV include a sudden reduction in food consumption, lethargy, loose cuticle, often reddish discolouration, and the presence of white spots on the inner surface of the carapace and cuticle over the abdominal segments (Takahashi et al. 1994). Transmission of the virus occurs mainly through oral ingestion and waterborne routes in farms and by vertical transmission in the case of shrimp hatcheries (Rosenberry 2003). White spot syndrome virus spreads through cannibalism of sick or dying shrimp or through contaminated water. Birds can carry infected shrimp between ponds. The virus can survive and remain infective in seawater for 4–7 days without a host.

To date, measures to control WSSV have included improving environmental rearing conditions and management practices (Rahman et al. 2007), using specialized formulated diets to boost shrimps' immune systems (Rajkumar et al. 2017), as well as using vaccines (Rijiravanich et al. 2008).

### 9.6.1 *Infectious Hypodermal and Hematopoietic Necrosis Virus (IHHNV)*

Infectious hypodermal and hematopoietic necrosis virus (IHHNV) is a widely distributed single-stranded DNA parvovirus that has been responsible for major losses in wild and farmed penaeid shrimp populations along the northwestern Pacific coast of Mexico since the early 1990s (Robles-Sikisaka et al. 2010). IHHNV was first described as the cause of acute epizootics and mass mortalities (> 90%) in juvenile and sub-adult *L. stylirostris* farmed in super-intensive raceway systems in Hawaii (Brock 1983; Lightner et al. 1983b, 1983c). Natural infections have been reported from *P. stylirostris*, *P. vannamei*, *P. occidentalis*, *P. californiensis*, *P. monodon*, *P. semisulcatus*, and *P. japonicus*. *P. setiferus*, *P. dourarum*, and *P. aztecus* have been infected experimentally with IHHNV, and *P. indicus and P. merguiensis* appear to be refractory to infection (Brock and Lightner 1990; Lightner 1996a).

The clinical signs of IHHNV disease in *P. stylirostris* are nonspecific and include anorexia, lethargy, and erratic swimming. Infected shrimps have been observed to rise to the surface of the water, remain motionless for a few moments, then roll over and sink to the bottom. Mortality may exceed 90% within several weeks of the onset of infection in juvenile *P. stylirostris* (Bell and Lightner, 1987). The horizontal transmission is carried out by cannibalism and through contaminated water (Lightner 1996b; Tang et al. 2003) and vertical transmission via infected eggs (Motte et al. 2003). IHHNV is closely related to densoviruses of the genus Brevidensovirus (Shike et al. 2000). In India, molecular evidence of the presence of IHHNV in *P. monodon* showing slow growth has been observed (Rai et al. 2009). Although there is no effective treatment to cure viral infections in crustaceans, biosecurity measures such as maintaining better water quality, selecting disease-free brood stocks or seeds, augmenting the disease resistance of the host, and hindering the disease transmission process are the major preventative measures that should be ensured to prevent the spread of these diseases.

### 9.6.2 *Hepatopancreatic Parvo-like Virus (HPV)*

*Penaeus monodon* densovirus (*Pm*DNV) (formerly hepatopancreatic parvovirus, or HPV) of penaeid shrimp is one of the important shrimp viruses that causes considerable economic loss in shrimp culture all over the world (Flegel 2006). *Pm*DNV, a viral pathogen, belongs to the family Parvoviridae (Bonami et al. 1995). HPV was identified simultaneously in cultured populations of *Penaeus semisulcatus* and *Penaeus merguiensis* in Asia (Lightner and Redman 1985). HPV has been documented in six other species, namely, *P. monodon, P. esculentus, P. indicus, P. chinensis, P. penicillatus*, and *P. vannamei*. The signs of disease in individual shrimp are not specific to HPV and include reduced growth, reduced preening, muscle opacity, and hepatopancreatic atrophy. Cumulative HPV-associated mortality was reported to be 50–100% after 4–8 weeks in juvenile *P. merguiensis* (Lightner and Redman 1985). HPV infects the epithelial cells of the hepatopancreas. The mode of transmission of HPV is not fully understood, as it has not been transmitted experimentally. Evidence exists that HPV is transmitted vertically from brood stock to progeny and horizontally during the post-larval stages (Brock and Lightner 1990).

HPV in penaeid shrimp causes considerable economic loss in shrimp culture all over the world (Flegel 2006). HPV is highly prevalent in some areas; it was found in 46% of the wild shrimp populations in India (Manjanaik et al. 2005). For farmed *P. monodon*, 31–62% of shrimp were found to be HPV-positive in India and Thailand (Umesha et al. 2003; Flegel et al. 2004). HPV has been found in nearly 100% of the populations of wild *P. monodon* in Africa and also in the wild stock of *P. merguiensis* in New Caledonia (Tang et al. 2008). Two completely sequenced genomes of HPV are available from Thailand (*P. monodon* densovirus) and Australia (*P. merguiensis* densovirus) (Sukhumsirichart et al. 2006; La Fauce et al. 2007). The complete nucleic acid sequence of *Pm*DNV from India revealed that the Indian *Pm*DNV is more closely related to Thai isolates than the other parvoviruses (Safeena et al. 2010).

HPV is a non-enveloped icosahedral virus with an average diameter of 22–24 nm. It has a minus genome with single-stranded DNA due to its distinct genome structure and an approximate genome size of 6 kb. HPV is considered a new member of Densovirinae, a subfamily which is able to infect both vertebrates and invertebrates. HPV-infected penaeid shrimp shows morphological differences in the appearance of the viral inclusion as well as variations in tissue tropism. Hepatopancreatic parvovirus (HPV) infects the hepatopancreas in penaeid shrimp and retards their growth (Phromjai et al. 2002). HPV intranuclear inclusions can be frequently observed in the midgut mucosal epithelium of HPV-infected *P. monodon* post-larvae from Madagascar (Pantoja and Lightner 2000). HPV has a marked tropism for epithelial cells of the hepatopancreas. It was hypothesized that infected shrimp could continuously shed viral particles, almost directly into the feces, after the host cells die and lyse. Catap and Travina (Catap and Traviña 2005) revealed that no HPV infection was produced in adult *P. monodon.*

HPV continues to cause substantial losses in the aquaculture industry in many countries. Control measures such as improvement of environmental conditions, stocking of specific pathogen-free shrimp post-larvae, and augmentation of disease resistance by vaccination using recombinant and RNAi techniques to prevent HPV infection in shrimp are in the experimental stages and are being applied to control HPV infection in the culture systems.

### 9.6.3 White Tail Disease (WTD)

White tail disease (WTD) is an important viral infection for *M. rosenbergii* due to large-scale mortalities in hatcheries and nurseries, leading to subsequent production losses in many countries such as Taiwan, Thailand, France, India, and the People's

Republic of China (Bonami and Widada 2011). In India, these viral pathogens have been reported in hatcheries and nursery ponds located in Andhra Pradesh and Tamil Nadu (India) (Hameed et al. 2004). White tail disease was also reported in China (Qian et al. 2003), Chinese Taipei (Wang et al. 2006), Thailand (Yoganandhan et al. 2006), Australia (Owens et al. 2009), and Malaysia (Saedi et al. 2012). The causative organisms for WTD were found to be two viral pathogens, namely, *M. rosenbergii* nodavirus (*Mr*NV) and extra-small virus (XSV). The typical gross signs of WTD in infected PL were whitish coloration of muscles, starting in some areas of the tail, extending to the tail muscles (abdomen), and at a final stage to all the muscles of the prawn, including the head (cephalothorax) muscles, and causing lethargy, abnormal behaviour, anorexia, and weakening of their feeding and swimming abilities. Degeneration of the telson and uropods is observed in severe cases. In all cases, mortality reached 100% within 2–3 days after the first appearance of prawns with whitish muscles (Arcier et al. 1999; Hameed et al. 2004; Sudhakaran et al. 2007). When investigated by histology, lesions were evidenced essentially in muscle and connective tissues. There are basophilic cytoplasmic inclusions with a diameter of 1–40 μm in the striated muscles of the abdomen, cephalothorax, and intratubular connective tissue of the hepatopancreas. No viral inclusions were observed in epithelial cells of the hepatopancreatic tubules or in midgut mucosal epithelial cells (Arcier et al. 1999).

*Mr*NV is a small, icosahedral, non-enveloped particle, measuring 26–27 nm in diameter. It contains two single-stranded RNAs (RNA1–2.9 kb and RNA2-1.26 kb). Its capsid contains a single polypeptide of 43 kDa. With these characteristics, it is closely related to the *Nodaviridae* family. Later, a second viruslike particle, unusually small (15 nm in diameter) and consequently named XSV (extra small viruslike particles), was also found associated with *Mr*NV (Qian et al. 2003). XSV is also a non-enveloped icosahedral virus with a linear ssRNA genome of 0.9 kb encoding two overlapping structural proteins of 16 and 17 kDa (Sri Widada and Bonami 2004; Bonami et al. 2005). Because of its small size and absence of gene-encoding enzymes required for replication, it has been suggested that it is a satellite virus or helper virus for *Mr*NV. The nodaviruses are known to contain a genome consisting of two single-stranded positive-sense RNA segments: RNA1, which encodes the viral part of the RNA-dependent RNA polymerase (RdRp), and RNA2, which encodes the capsid protein gene of 43 kDa (Bonami et al. 2005). The genome of XSV consists of a linear single-stranded positive-sense RNA coding for a capsid protein gene of 17 kDa (capsid protein-17). Because of its extremely small size and absence of gene-encoding enzymes required for replication, it has been suggested that XSV may be a satellite virus, while *Mr*NV plays the role of a helper virus (Sri Widada and Bonami 2004). Nucleotide sequencing of the *Mr*NV genome indicated that RNA-1 was composed of 3.2 kb nucleotides and that RNA-2 contained 1.17 kb nucleotides.

In horizontal transmission experiments, five developmental stages of *Artemia* were exposed to *Mr*NV and XSV via immersion and oral routes, and it was reported that *Artemia* is capable of transmitting these viruses to *M. rosenbergii* post-larvae and is responsible for causing WTD (Sudhakaran et al. 2006a, b). WTD mainly affects the PL of *M. rosenbergii* and causes major mortalities in these young animals. Prawn brood stock inoculated with *Mr*NV and XSV by oral or immersion challenge survived without clinical signs of WTD. The survival rate of larvae gradually decreased, and 100% mortality was observed in the post-larvae. Experimental infection of brooders with both *Mr*NV and XSV demonstrated that the vertical route is actually the main mechanism of disease transmission. In the infected brooders, ovarian tissue and fertilized eggs were found to be positive for *Mr*NV/XSV as evidenced by RT-PCR, and a mortality of up to 100% was observed in PL from hatched eggs released from virus-inoculated brooders (Sudhakaran et al. 2007). Control measures such as improvement of environmental conditions, stocking of specific pathogen-free post-larvae, augmentation of disease resistance by immunostimulants, and vaccination using recombinant and RNAi techniques to prevent *Mr*NV/XSV infection.

### *9.6.4 Infectious Myonecrosis Virus (IMNV)*

Infectious myonecrosis (IMNV) is an emerging OIE-listed potential shrimp virus that can cause significant losses in shrimp aquaculture (Nunes et al. 2004). The disease was first reported in cultured *L. vannamei* in Brazil in 2003 (Tang et al. 2005) and later in Indonesia (Senapin et al. 2007) and most recently in India (Sahul Hameed et al. 2017). IMNV belongs to the family Totiviridae and is closely related to *Giardia lamblia* virus. It measures 40 nm in diameter and has an unenveloped icosahedral shape (Lightner et al. 2004; Fauquet et al. 2005; Lightner 2011). The viral genome consists of a single double-stranded RNA of 7,561–8,230 bp (Dantas et al. 2015; Naim et al., 2015; Senapin et al. 2007).

The susceptible hosts for the virus are brown tiger prawn (*Penaeus esculentus*), banana prawn (*P. merguiensis*), and whiteleg shrimp (*P. vannamei*) (OIE 2021). Pathogenicity of this virus, which ranged from 40% to 70% mortality in cultured *L. vannamei*, was reported (Andrade et al. 2008). No other hosts susceptible to IMNV have been reported so far. Clinical signs of the disease are characterized by opaque or discolored skeletal muscle tissues, primarily in the abdominal segment of affected shrimp (Poulos et al. 2006). Histopathology shows that muscle lesions are characterized by myonecrosis, hemocyte infiltration, and the presence of basophilic, cytoplasmic inclusions (Poulos and Lightner 2006; Poulos et al. 2006). The virus shows both horizontal and vertical modes of transmission in *P. vannamei* (Lightner et al. 2004; da Silva et al. 2016).

Control measures to prevent the spread of IMNV infection are good management practices of shrimp farming, unregulated transboundary movement of brood stocks, and post-larvae.

## References


Abbas WT (2021) Advantages and prospective challenges of nanotechnology applications in fish cultures: a comparative review. Environ Sci Pollut Res 28(7):7669–7690. https://doi.org/10.1007/s11356-020-12166-0

Abdullah A, Ramly R, Ridzwan MM, Sudirwan F, Abas A, Ahmad K, Murni M, Kua BC (2018) First detection of tilapia lake virus (TiLV) in wild river carp (Barbonymus schwanenfeldii) at Timah Tasoh Lake. Malaysia J Fish Dis 41(9):1459–1462

Abou El-Nour KMM, Eftaiha AA, Al-Warthan A, Ammar RAA (2010) Synthesis and applications of silver nanoparticles. Arab J Chem 3(3):135–140. https://doi.org/10.1016/j.arabjc.2010.04.008

Adamek M, Oschilewski A, Wohlsein P, Jung-Schroers V, Teitge F, Dawson A, Gela D, Piackova V, Kocour M, Adamek J, Bergmann SM (2017) Experimental infections of different carp strains with the carp edema virus (CEV) give insights into the infection biology of the virus and indicate possible solutions to problems caused by koi sleepy disease (KSD) in carp aquaculture. Vet Res 48(1):1–16

Adams A (2019) Progress, challenges and opportunities in fish vaccine development. Fish & Shellfish Immunol 90:210–214. https://doi.org/10.1016/j.fsi.2019.04.066

Adomako M et al (2012) Oral DNA vaccination of rainbow trout, Oncorhynchus mykiss (Walbaum), against infectious haematopoietic necrosis virus using PLGA [Poly(D,L-Lactic-Co-Glycolic Acid)] nanoparticles. J Fish Dis 35(3):203–214. https://doi.org/10.1111/j.1365-2761.2011.01338.x

Amal MNA, Koh CB, Nurliyana M, Suhaiba M, Nor-Amalina Z, Santha S, Diyana-Nadhirah KP, Yusof MT, Ina-Salwany MY, Zamri-Saad M (2018) A case of natural co-infection of Tilapia Lake Virus and Aeromonas veronii in a Malaysian red hybrid tilapia (Oreochromis niloticus× O. mossambicus) farm experiencing high mortality. Aquaculture 485:12–16

Andrade TP, Redman RM, Lightner DV (2008) Evaluation of the preservation of shrimp samples with Davidson's AFA fixative for infectious myonecrosis virus (IMNV) in situ hybridization. Aquaculture 278(1-4):179–183

Arcier JM, Herman F, Lightner DV, Redman RM, Mari J, Bonami JR (1999) A viral disease associated with mortalities in hatchery-reared postlarvae of the giant freshwater prawn Macrobrachium rosenbergii. Dis Aquat Org 38(3):177–181

Arimoto M, Mushiake K, Mizuta Y, Nakai T, Muroge K, Furusawa I (1992) Detection of striped jack nervous necrosis virus (SJNNV) by enzyme-linked immunosorbent assay (ELISA). Fish Pathol 27(4):191–195

Azad IS, Shekhar MS, Thirunavukkarasu AR, Poornima M, Kailasam M, Rajan JJS, Ali SA, Abraham M, Ravichandran P (2005) Nodavirus infection causes mortalities in hatchery produced larvae of Lates calcarifer: first report from India. Dis Aquat Org 63(2-3):113–118

Bacharach E, Mishra N, Briese T, Zody MC, Kembou Tsofack JE, Zamostiano R, Berkowitz A, Ng J, Nitido A, Corvelo A, Toussaint NC (2016) Characterization of a novel orthomyxo-like virus causing mass die-offs of tilapia. MBio 7(2):e00431–e00416

Becker JA, Tweedie A, Rimmer A, Landos M, Lintermans M, Whittington RJ (2014) Incursions of Cyprinid herpesvirus 2 in goldfish populations in Australia despite quarantine practices. Aquaculture 432:53–59

Bedekar MK, Kole S (2022) Development of Nano-Conjugated DNA Vaccine Against Edwardsiellosis Disease in Fish. Methods Mol Biol (Clifton, NJ) 2411:195–204. https://doi.org/10.1007/978-1-0716-1888-2_11

Behera BK, Pradhan PK, Swaminathan TR, Sood N, Paria P, Das A, Verma DK, Kumar R, Yadav MK, Dev AK, Parida PK (2018) Emergence of tilapia lake virus associated with mortalities of farmed Nile tilapia Oreochromis niloticus (Linnaeus 1758) in India. Aquaculture 484:168–174

Belcher CR, Young PR (1998) Colourimetric PCR-based detection of monodon baculovirus in whole Penaeus monodon postlarvae. J Virol Methods 74(1):21–29

Belton B, Thilsted SH (2014) Fisheries in transition: Food and nutrition security implications for the global South. Glob Food Sec 3(1):59–66

Belton B, Little DC, Zhang W, Edwards P, Skladany M, Thilsted SH (2020) Farming fish in the sea will not nourish the world. Nat Commun 11(1):1–8

Béné C, Arthur R, Norbury H, Allison EH, Beveridge M, Bush S, Campling L, Leschen W, Little D, Squires D, Thilsted SH (2016) Contribution of fisheries and aquaculture to food security and poverty reduction: assessing the current evidence. World Dev 79:177–196

Boitard PM, Baud M, Labrut S, de Boisséson C, Jamin M, Bigarré L (2016) First detection of Cyprinid Herpesvirus 2 (Cy HV-2) in goldfish (Carassius auratus) in France. J Fish Dis 39(6):673–680

Bonami JR, Widada JS (2011) Viral diseases of the giant fresh water prawn Macrobrachium rosenbergii: a review. J Invertebr Pathol 106(1):131–142

Bonami JR, Mari J, Poulos BT, Lightner DV (1995) Characterization of hepatopancreatic parvo-like virus, a second unusual parvovirus pathogenic for penaeid shrimps. J Gen Virol 76(4):813–817

Bonami JR, Shi Z, Qian D, Sri Widada J (2005) White tail disease of the giant freshwater prawn, Macrobrachium rosenbergii: separation of the associated virions and characterization of MrNV as a new type of nodavirus. J Fish Dis 28(1):23–31

Bondad-Reantaso MG, Subasinghe RP, Arthur JR, Ogawa K, Chinabut S, Adlard R, Tan Z, Shariff M (2005) Disease and health management in Asian aquaculture. Vet Parasitol 132(3-4):249–272

Breuil B, Pepin JFP, Boscher S, Thiery R (2002) Experimental vertical transmission of nodavirus from broodfish to eggs and larvae of the sea bass, Dicentrarchus labrax (L.). J Fish Dis 25:697–702

Brock JA, Lightner DV (1990) Diseases of crustacean. In: Disease of Marine Animals, vol 3. Biologische Anstalt Helgoland, Kinne, Humberg, pp 245–424

Brock JA (1983) Diseases (infectious and non-infectious), metazoan parasites, predators and public health considerations in Macrobrachium culture and fisheries. CRC Handbook Maricult 1:329–370

Castric J, Thiery R, Jeffroy J, De Kinkelin P, Raymond JC (2001) Sea bream Sparus aurata, an asymptomatic contagious fish host for nodavirus. Dis Aquat Org 47(1):33–38

Catap, E.S. and Traviña, R.D., 2005. Experimental transmission of hepatopancreatic parvovirus (HPV) infection in Penaeus monodon postlarvae. In Fifth Symposium on Symposium on Diseases in Asian Aquaculture, held in Gold Coast, Australia on 24-28 November 2002 (pp. 415-420). Fish Health Section, Asian Fisheries Society.

Chen SN, Lo CF, Lui SM, Kou GH (1989) 'The first identification of Penaeus monodon baculovirus (MBV) in cultured sand shrimp, Metapenaeus ensis. Bull Eur Assoc Fish Pathol 9:62–64

Costa JZ, Thompson KD (2016) Understanding the interaction between Betanodavirus and its host for the development of prophylactic measures for viral encephalopathy and retinopathy. Fish Shellfish Immunol 53:35–49

da Silva SMBC, da Silva ADR, Lavander HD, Chaves TCB, Peixoto S, Gálvez AO, Coimbra MRM (2016) Vertical transmission of infectious myonecrosis virus in Litopenaeus vannamei. Aquaculture 459:216–222

Dantas MD, Chavante SF, Teixeira DI, Lima JP, Lanza DC (2015) Analysis of new isolates reveals new genome organization and a hypervariable region in infectious myonecrosis virus (IMNV). Virus research 203:66–71

Davison AJ, Kurobe T, Gatherer D, Cunningham C, Korf I, Fukuda H, Hedrick RP, Waltzek TB (2013) Comparative genomics of carp herpesviruses. J Virol 87(5):2908–2922

Del-Pozo J, Mishra N, Kabuusu R, Cheetham S, Eldar A, Bacharach E, Lipkin WI, Ferguson HW (2017) Syncytial hepatitis of tilapia (Oreochromis niloticus L.) is associated with orthomyxovirus-like virions in hepatocytes. Vet Pathol 54(1):164–170

Dong HT, Senapin S, Gangnonngiw W, Nguyen VV, Rodkhum C, Debnath PP, Delamare-Deboutteville J, Mohan CV (2020) Experimental infection reveals transmission of tilapia lake virus (TiLV) from tilapia broodstock to their reproductive organs and fertilized eggs. Aquaculture 515:734541

Doszpoly A, Benko M, Csaba G, Dan A, Lang M, Harrach B (2011) Introduction of the family Alloherpesviridae: the first molecular detection of herpesviruses of cyprinid fish in Hungary. Magyar Allatorvosok Lapja 133(3):174–181

Embregts CWE, Forlenza M (2016) Oral vaccination of fish: Lessons from humans and veterinary species. Dev Comp Immunol 64:118–137. https://doi.org/10.1016/j.dci.2016.03.024

Escobedo-Bonilla CM, Alday-Sanz V, Wille M, Sorgeloos P, Pensaert MB, Nauwynck HJ (2008) A review on the morphology, molecular characterization, morphogenesis and pathogenesis of white spot syndrome virus. J Fish Dis 31(1):1–18

Eyngor M, Zamostiano R, Kembou Tsofack JE, Berkowitz A, Bercovier H, Tinman S, Lev M, Hurvitz A, Galeotti M, Bacharach E, Eldar A (2014) Identification of a novel RNA virus lethal to tilapia. J Clin Microbiol 52(12):4137–4146

Fathi M, Dickson C, Dickson M, Leschen W, Baily J, Muir F, Ulrich K, Weidmann M (2017) Identification of Tilapia Lake Virus in Egypt in Nile tilapia affected by 'summer mortality'syndrome. Aquaculture 473:430–432

Fauquet CM, Mayo MA, Maniloff J, Desselberger U, Ball LA (2005) Totiviridae. In: Virus taxonomy: Classification and nomenclature of viruses. Elsevier, San Francisco, CA, pp 571–580). Eighth Report of the International Committee on the Taxonomy of Viruses

Ferguson HW, Kabuusu R, Beltran S, Reyes E, Lince JA, del Pozo J (2014) Syncytial hepatitis of farmed tilapia, Oreochromis niloticus (L.): a case report. J Fish Dis 37(6):583–589

Fichi G, Susini F, Cocumelli C, Cersini A, Salvadori M, Guarducci M, Cardeti G (2016) New detection of Cyprinid herpesvirus 2 in mass mortality event of Carassius carassius (L.), in Italy. J Fish Dis 39(12):1523–1527

Flegel TW (1997) Major viral diseases of the black tiger prawn (Penaeus monodon) in Thailand. World J Microbiol Biotechnol 13(4):433–442

Flegel TW (2006) Detection of major penaeid shrimp viruses in Asia, a historical perspective with emphasis on Thailand. Aquaculture 258(1-4):1–33

Flegel TW, Nielsen L, Thamavit V, Kongtim S, Pasharawipas T (2004) Presence of multiple viruses in non-diseased, cultivated shrimp at harvest. Aquaculture 240(1-4):55–68

Gangnonngiw W, Laisutisan K, Sriurairatana S, Senapin S, Chuchird N, Limsuwan C, Chaivisuthangkura P, Flegel TW (2010) Monodon baculovirus (MBV) infects the freshwater prawn Macrobrachium rosenbergii cultivated in Thailand. Virus Res 148(1-2):24–30

Girisha SK, Kushala KB, Nithin MS, Puneeth TG, Naveen Kumar BT, Vinay TN, Suresh T, Ajay SK, Venugopal MN, Ramesh KS (2021) First report of the infectious spleen and kidney necrosis virus (ISKNV) infection in ornamental fishes in India. Transbound Emerg Dis 68(2):964–972

Girisha SK, Puneeth TG, Nithin MS, Kumar BN, Ajay SK, Vinay TN, Suresh T, Venugopal MN, Ramesh KS (2020) Red sea bream iridovirus disease (RSIVD) outbreak in Asian seabass (Lates calcarifer) cultured in open estuarine cages along the west coast of India: First report. Aquaculture 520:734712

Go J, Whittington R (2006) Experimental transmission and virulence of a megalocytivirus (Family Iridoviridae) of dwarf gourami (Colisa lalia) from Asia in Murray cod (Maccullochella peelii peelii) in Australia. Aquaculture 258(1-4):140–149

Goodwin AE, Khoo L, LaPatra SE, Bonar C, Key DW, Garner M, Lee MV, Hanson L (2006) Goldfish hematopoietic necrosis herpesvirus (cyprinid herpesvirus 2) in the USA: molecular confirmation of isolates from diseased fish. J Aquat Anim Health 18(1):11–18

Goodwin AE, Sadler J, Merry GE, Marecaux EN (2009) Herpesviral haematopoietic necrosis virus (CyHV-2) infection: case studies from commercial goldfish farms. J Fish Dis 32(3):271–278

Grotmol S, Totland GK (2000) Surface disinfection of Atlantic halibut Hippoglossus hippoglossus eggs with ozonated sea-water inactivates nodavirus and increases survival of the larvae. Dis Aquat Org 39(2):89–96

Grotmol S, Bergh Ø, Totland GK (1999) Transmission of viral encephalopathy and retinopathy (VER) to yolk-sac larvae of the Atlantic halibut Hippoglossus hippoglossus: occurrence of nodavirus in various organs and a possible route of infection. Dis Aquat Org 36(2):95–106

Hameed AS, Yoganandhan K, Widada JS, Bonami JR (2004) Studies on the occurrence of Macrobrachium rosenbergii nodavirus and extra small virus-like particles associated with white tail disease of M. rosenbergii in India by RT-PCR detection. Aquaculture 238(1-4):127–133
Hanson L, Dishon A, Kotler M (2011) Herpesviruses that infect fish. Viruses 3(11):2160–2191
Hao Z et al (2016) Establishment of droplet digital PCR assay for the rapid detection of cyprinid herpesvirus 2. Chinese Veterinary Science/Zhongguo Shouyi Kexue 46(2):167–173
He JG, Deng M, Weng SP, Li Z, Zhou SY, Long QX, Wang XZ, Chan SM (2001) Complete genome analysis of the mandarin fish infectious spleen and kidney necrosis iridovirus. Virology 291(1):126–139
He JG, Zeng K, Weng SP, Chan SM (2002) Experimental transmission, pathogenicity and physical–chemical properties of infectious spleen and kidney necrosis virus (ISKNV). Aquaculture 204(1-2):11–24
Heyerdahl LW et al (2018) Innovative vaccine delivery strategies in response to a cholera outbreak in the challenging context of Lake Chilwa. A rapid qualitative assessment. Vaccine 36(44):6491–6496. https://doi.org/10.1016/j.vaccine.2017.10.108
Hu Q, Li H, Wang L, Gu H, Fan C (2018) DNA nanotechnology-enabled drug delivery systems. Chem Rev 119(10):6459–6506
Huang X et al (2021) Oral Probiotic Vaccine Expressing Koi Herpesvirus (KHV) ORF81 Protein Delivered by Chitosan-Alginate Capsules Is a Promising Strategy for Mass Oral Vaccination of Carps against KHV Infection. J Virol 95(12):e00415. https://doi.org/10.1128/jvi.00415-21
ICTV, 2018. Virus Taxonomy: 2018b Release. International Committee on Taxonomy of Viruses. https://talk.ictvonline.org/taxonomy
Jaemwimol P, Rawiwan P, Tattiyapong P, Saengnual P, Kamlangdee A, Surachetpong W (2018) Susceptibility of important warm water fish species to tilapia lake virus (TiLV) infection. Aquaculture 497:462–468
Jansen MD, Dong HT, Mohan CV (2019) Tilapia lake virus: a threat to the global tilapia industry? Rev Aquac 11(3):725–739
Jeffery KR, Bateman K, Bayley A, Feist SW, Hulland J, Longshaw C, Stone D, Woolford G, Way K (2007) Isolation of a cyprinid herpesvirus 2 from goldfish, Carassius auratus (L.), in the UK. J Fish Dis 30(11):649–656
Jeong KH, Kim HJ, Kim HJ (2020) Current status and future directions of fish vaccines employing virus-like particles. Fish Shellfish Immunol 100:49–57. https://doi.org/10.1016/j.fsi.2020.02.060
Jeong JB, Kim HY, Jun LJ, Lyu JH, Park NG, Kim JK, Do Jeong H (2008) Outbreaks and risks of infectious spleen and kidney necrosis virus disease in freshwater ornamental fishes. Dis Aquat Org 78(3):209–215
Ji J, Torrealba D, Ruyra À, Roher N (2015) Nanodelivery Systems as New Tools for Immunostimulant or Vaccine Administration: Targeting the Fish Immune System. Biology 4(4):664–696. https://doi.org/10.3390/biology4040664
Johnson PT, Lightner DV (1988) Rod-shaped nuclear viruses of crustaceans: gut-infecting species. Dis Aquat Org 5:123–141
Jung SJ, Miyazaki T (1995) Herpesviral haematopoietic necrosis of goldfish, Carassius auratus (L.). J Fish Dis 18(3):211–220
Jung-Schroers V, Adamek M, Wohlsein P, Wolter J, Wedekind H, Steinhagen D (2016) First outbreak of an infection with infectious spleen and kidney necrosis virus (ISKNV) in ornamental fish in Germany. Dis Aquat Org 119(3):239–244
Karunasagar I, Karunasagar I (1998) Disease problems affecting cultured penaeid shrimp in India. Fish Pathol 33(4):413–419
Kawakami H, Nakajima K (2002) Cultured fish species affected by red sea bream iridoviral disease from 1996 to 2000. Fish Pathol 37(1):45–47
Kayansamruaj P, Areechon N, Unajak S (2020) Development of fish vaccine in Southeast Asia: A challenge for the sustainability of SE Asia aquaculture. Fish Shellfish Immunol 103:73–87. https://doi.org/10.1016/j.fsi.2020.04.031

Kennedy DA, Kurath G, Brito IL, Purcell MK, Read AF, Winton JR, Wargo AR (2016) Potential drivers of virulence evolution in aquaculture. Evol Appl 9(2):344–354

Kole S, Qadiri SSN, Shin SM, Kim WS, Lee J, Jung SJ (2019a) Nanoencapsulation of inactivated-viral vaccine using chitosan nanoparticles: Evaluation of its protective efficacy and immune modulatory effects in olive flounder (Paralichthys olivaceus) against viral haemorrhagic septicaemia virus (VHSV) infection. Fish Shellfish Immunol 91:136–147. https://doi.org/10.1016/j.fsi.2019.05.017

Kole S, Qadiri SSN, Shin SM, Kim WS, Lee J, Jung SJ (2019b) PLGA encapsulated inactivated-viral vaccine: Formulation and evaluation of its protective efficacy against viral haemorrhagic septicaemia virus (VHSV) infection in olive flounder (Paralichthys olivaceus) vaccinated by mucosal delivery routes. Vaccine 37(7):973–983. https://doi.org/10.1016/j.vaccine.2018.12.063

Kumar P, Khar S, Dwivedi S, Sharma SK, Sharma H (2015) An overview of fisheries and aquaculture in India. Agro Economist Int J 2(2):1

Kurita J, Nakajima K (2012) Megalocytiviruses. Viruses 4(4):521–538

La Fauce KA, Elliman J, Owens L (2007) Molecular characterisation of hepatopancreatic parvovirus (PmergDNV) from Australian Penaeus merguiensis. Virology 362(2):397–403

Leobert D, Lavilla-Pitogo CR, Villar CBR, Paner MG, Capulos GC (2008) Prevalence of monodon baculovirus (MBV) in wild shrimp Penaeus monodon in the Philippines. Aquaculture 285(1-4):19–22

Lewisch E, Gorgoglione B, Way K, El-Matbouli M (2015) Carp edema virus/Koi sleepy disease: An emerging disease in Central-East Europe. Transbound Emerg Dis 62(1):6–12

Li L, Lin SL, Deng L, Liu ZG (2013) Potential use of chitosan nanoparticles for oral delivery of DNA vaccine in black seabream Acanthopagrus schlegelii Bleeker to protect from Vibrio parahaemolyticus. J Fish Dis 36(12):987–995. https://doi.org/10.1111/jfd.12032

Liamnimitr P, Thammatorn W, Sonicha U, Tattiyapong P, Surachetpong W (2018) Non-lethal sampling for Tilapia Lake Virus detection by RT-qPCR and cell culture. Aquaculture 486:75–80

Lightner DV, Redman RM (1981) A baculovirus-caused disease of the penaeid shrimp, Penaeus monodon. J Invertebr Pathol 38(2):299–302

Lightner DV, Redman RM (1985) A parvo-like virus disease of penaeid shrimp. J Invertebr Pathol 45(1):47–53

Lightner DV, Redman RM (1998) Shrimp diseases and current diagnostic methods. Aquaculture 164(1-4):201–220

Lightner, D.V., 1996a. A handbook of shrimp pathology and diagnostic procedures for diseases of cultured penaeid shrimp.

Lightner DV (1996b) Epizootiology, distribution and the impact on international trade of two penaeid shrimp viruses in the Americas. Revue scientifique et technique (International Office of Epizootics) 15(2):579–601

Lightner DV (2011) Virus diseases of farmed shrimp in the Western Hemisphere (the Americas): a review. J Invertebr Pathol 106(1):110–130

Lightner DV, Pantoja CR, Poulos BT, Tang KFJ, Redman RM, Andreas T, Bonami JR (2004) Infectious myonecrosis (IMN): a new virus disease of Litopenaeus vannamei. Aquaculture 242:353

Lightner DV, Redman RM, Bell TA (1983a) Observations on the geographic distribution, pathogenesis and morphology of the baculovirus from Penaeus monodon Fabricius. Aquaculture 32(3-4):209–233

Lightner DV, Redman RM, Bell TA (1983c) Infectious hypodermal and hematopoietic necrosis, a newly recognized virus disease of penaeid shrimp. J Invertebr Pathol 42(1):62–70

Lightner DV, Redman RM, Bell TA, Brock JA (1983b) Detection of IHHN virus in Penaeus stylirostris and P. vannamei imported into Hawaii. J World Maricult Soc 14(1-4):212–225

Luis AIS et al (2021) Ecotoxicity evaluation of polymeric nanoparticles loaded with ascorbic acid for fish nutrition in aquaculture. J Nanobiotechnol 19(1):163. https://doi.org/10.1186/s12951-021-00910-8

Luo D, Liang L, Xie J, Wu X (2014) Development of duplex PCR for detection of Cyprinid herpesvirus 2. Zhongguo Yufang Shouyi Xuebao/Chinese Journal of Preventive Veterinary Medicine 36(5):379–382

Manivannan S, Kennedy B, Karunasagar I, Karunasagar I (2004) Prevalence of monodon baculovirus in wild Metapenaeus species along the southwest coast of India. Aquaculture 232(1-4):63–67

Manjanaik B, Umesha KR, Karunasagar I, Karunasagar I (2005) Detection of hepatopancreatic parvovirus (HPV) in wild shrimp from India by nested polymerase chain reaction (PCR). Dis Aquat Org 63(2-3):255–259

Matras M, Stachnik M, Borzym E, Maj-Paluch J, Reichert M (2019) Potential vector species of carp edema virus (CEV). J Fish Dis 42(7):959–964

Matsuoka S, Inouye K, Nakajima K (1996) Cultured fish species affected by red sea bream iridoviral disease from 1991 to 1995. Fish Pathol 31(4):233–234

Miyazaki T, Isshiki T, Katsuyuki H (2005) Histopathological and electron microscopy studies on sleepy disease of koi Cyprinus carpio koi in Japan. Dis Aquat Org 65(3):197–207

Mori KI, Nakai T, Muroga K, Arimoto M, Mushiake K, Furusawa I (1992) Properties of a new virus belonging to Nodaviridae found in larval striped jack (Pseudocaranx dentex) with nervous necrosis. Virology 187(1):368–371

Motte E, Yugcha E, Luzardo J, Castro F, Leclercq G, Rodríguez, J., Miranda, P., Borja, O., Serrano, J., Terreros, M. and Montalvo, K. (2003) Prevention of IHHNV vertical transmission in the white shrimp Litopenaeus vannamei. Aquaculture 219(1-4):57–70

Mugimba KK, Chengula AA, Wamala S, Mwega ED, Kasanga CJ, Byarugaba DK, Mdegela RH, Tal S, Bornstein B, Dishon A, Mutoloki S (2018) Detection of tilapia lake virus (Ti LV) infection by PCR in farmed and wild Nile tilapia (Oreochromis niloticus) from Lake Victoria. J Fish Dis 41(8):1181–1189

Mugimba KK, Tal S, Dubey S, Mutoloki S, Dishon A, Evensen Ø, Munang'andu HM (2019) Gray (Oreochromis niloticus x O. aureus) and red (Oreochromis spp.) Tilapia show equal susceptibility and proinflammatory cytokine responses to experimental Tilapia Lake Virus infection. Viruses 11(10):893

Munday BL, Kwang J, Moody N (2002) Betanodavirus infections of teleost fish: a review. J Fish Dis 25(3):127–142

Murakami Y (1976) Studies on mass mortality of juvenile carp.-about mass mortality showing edema. Report of Research on Fish Disease.

Naim S, Tang KF, Yang M, Lightner DV, Nibert ML (2015) Extended genome sequences of penaeid shrimp infectious myonecrosis virus strains from Brazil and Indonesia. Archives of Virology 160:1579–1583

Nakajima K, Maeno Y, Honda A, Yokoyama K, Tooriyama T, Manabe S (1999) Effectiveness of a vaccine against red sea bream iridoviral disease in a field trial test. Dis Aquat Org 36(1):73–75

Nanthini R, Abdul Majeed S, Vimal S, Taju G, Sivakumar S, Santhosh Kumar S, Pillai D, Sneha KG, Rakesh CG, Sahul Hameed AS (2019) In vitro propagation of tilapia lake virus in cell lines developed from Oreochromis mossambicus. J Fish Dis 42(11):1543–1552

Nunes AJ, Martins PCC, Gesteira TCV (2004) Carcinicultura ameaçada Rev Panoram. Aquic 83:37–51

OIE (2003) Chapter 2.1.7 Viral encephalopathy and retinopathy. pp. 135-141. In: Manual of Diagnostic Tests for Aquatic Animals, 4th edn. Office International des Epizooties, Paris, 358 pp

OIE (2006) Manual of diagnostic tests for aquatic animals, 5th edn. Office International des Epizooties, Paris, 358p

OIE (2019) Manual of diagnostic tests for aquatic animals. Red Sea Bream Iridoviral Disease. Chapter 2.3.8.

OIE (2021) Manual of diagnostic tests for aquatic animals. Infection with Infectious myonecrosis virus. Chapter 2.2.5.

Oroojalian F et al (2020) Recent advances in nanotechnology-based drug delivery systems for the kidney. J Control Release 321:442–462. https://doi.org/10.1016/j.jconrel.2020.02.027

Owens L, La Fauce K, Juntunen K, Hayakijkosol O, Zeng C (2009) Macrobrachium rosenbergii nodavirus disease (white tail disease) in Australia. Dis Aquat Org 85(3):175–180

Oyamatsu, T., Matoyama, H., Yamamoto, K. and Fukuda, H., 1997. A trial for the detection of carp [Cyprinus carpio] edema virus by using polymerase Chain reaction. Suisanzoshoku (Japan)

Pantoja CR, Lightner DV (2000) A non-destructive method based on the polymerase chain reaction for detection of hepatopancreatic parvovirus (HPV) of penaeid shrimp. Dis Aquat Org 39(3):177–182

Paynter JL, Vickers JE, Lester RJG (1992) Experimental transmission of Penaeus monodon-type baculovirus (MBV). Diseases in Asian Aquaculture, Asian Fisheries Society, Manila, pp 97–100

Peducasse S, Castric J, Thiery R, Jeffroy J, LeVen A, Laurencin FB (1999) Comparative study of viral encephalopathy and retinopathy in juvenile sea bass Dicentrarchus labrax infected in different ways. Dis Aquat Org 36:11–20

Phromjai J, Boonsaeng V, Withyachumnarnkul B, Flegel TW (2002) Detection of hepatopancreatic parvovirus in Thai shrimp Penaeus monodon by in situ hybridization, dot blot hybridization and PCR amplification. Dis Aquat Org 51(3):227–232

Poulos B, Tang K, Pantoja C, Bonami JR, Lightner D (2006) Purification and characterization of infectious myonecrosis virus of penaeid shrimp. J Gen Virol 87:987–996

Poulos BT, Lightner DV (2006) Detection of infectious myonecrosis virus (IMNV) of penaeid shrimp by reverse-transcriptase polymerase chain reaction (RT-PCR). Dis Aquat Org 73(1):69–72

Poulos BT, Lightner DV, Trumper B, Bonami JR (1994) Monoclonal antibodies to a penaeid shrimp parvovirus, infectious hypodermal and hematopoietic necrosis virus (IHHNV). J Aquat Anim Health 6(2):149–154

Pulido LLH, Mora CM, Hung AL, Dong HT, Senapin S (2019) Tilapia lake virus (TiLV) from Peru is genetically close to the Israeli isolates. Aquaculture 510:61–65

Puneeth TG, Baliga P, Girisha SK, Shekar M, Nithin MS, Suresh T, Kumar BN (2021) Complete genome analysis of a red seabream iridovirus (RSIV) isolated from Asian seabass (Lates calcarifer) in India. Virus Res 291:198199

Qian D, Shi Z, Zhang S, Cao Z, Liu W, Li L, Xie Y, Cambournac I, Bonami JR (2003) Extra small virus-like particles (XSV) and nodavirus associated with whitish muscle disease in the giant freshwater prawn, Macrobrachium rosenbergii. J Fish Dis 26(9):521–527

Rahman MM, Corteel M, Wille M, Alday-Sanz V, Pensaert MB, Sorgeloos P, Nauwynck HJ (2007) The effect of raising water temperature to 33á C in Penaeus vannamei juveniles at different stages of infection with white spot syndrome virus (WSSV). Aquaculture 272(1-4):240–245

Rai P, Pradeep B, Karunasagar I, Karunasagar I (2009) Detection of viruses in Penaeus monodon from India showing signs of slow growth syndrome. Aquaculture 289(3-4):231–235

Rajkumar T, Taju G, Majeed SA, Sajid MS, Kumar SS, Sivakumar S, Thamizhvanan S, Vimal S, Hameed AS (2017) Ontogenetic changes in the expression of immune related genes in response to immunostimulants and resistance against white spot syndrome virus in Litopenaeus vannamei. Dev Comp Immunol 76:132–142

Ramasamy P, Brennan GP, Jayakumar R (1995) A record and prevalence of monodon baculovirus from postlarval Penaeus monodon in Madras, India. Aquaculture 130(2-3):129–135

Rehman T, Yin L, Latif MB, Zhou Y, Wang K, Geng Y, Huang X, Chen D, Fang J, Chen Z, Guo H (2020) Current findings on carp edema virus, control challenges, and future outlook. Aquac Int 28(5):2015–2026

Rijiravanich A, Browdy CL, Withyachumnarnkul B (2008) Knocking down caspase-3 by RNAi reduces mortality in Pacific white shrimp Penaeus (Litopenaeus) vannamei challenged with a low dose of white-spot syndrome virus. Fish Shellfish Immunol 24(3):308–313

Rimmer AE, Becker JA, Tweedie A, Lintermans M, Landos M, Stephens F, Whittington RJ (2015) Detection of dwarf gourami iridovirus (Infectious spleen and kidney necrosis virus) in populations of ornamental fish prior to and after importation into Australia, with the first

evidence of infection in domestically farmed Platy (Xiphophorus maculatus). Prev Vet Med 122(1-2):181–194
Robles-Sikisaka R, Bohonak AJ, McClenaghan LR Jr, Dhar AK (2010) Genetic signature of rapid IHHNV (infectious hypodermal and hematopoietic necrosis virus) expansion in wild penaeus shrimp populations. PLoS One 5(7):e11799
Rosenberry, B., 2003. World shrimp farming 2003. Shrimps news international.
Saedi TA, Moeini H, Tan WS, Yusoff K, Daud HM, Chu KB, Tan SG, Bhassu S (2012) Detection and phylogenetic profiling of nodavirus associated with white tail disease in Malaysian Macrobrachium rosenbergii de Man. Mol Biol Rep 39(5):5785–5790
Safeena MP, Tyagi A, Rai P, Karunasagar I, Karunasagar I (2010) Complete nucleic acid sequence of Penaeus monodon densovirus (PmDNV) from India. Virus Res 150(1-2):1–11
Sahoo PK, Swaminathan TR, Abraham TJ, Kumar R, Pattanayak S, Mohapatra A, Rath SS, Patra A, Adikesavalu H, Sood N, Pradhan PK (2016) Detection of goldfish haematopoietic necrosis herpes virus (Cyprinid herpesvirus-2) with multi-drug resistant Aeromonas hydrophila infection in goldfish: First evidence of any viral disease outbreak in ornamental freshwater aquaculture farms in India. Acta Trop 161:8–17
Sahul Hameed AS, Abdul Majeed S, Vimal S, Madan N, Rajkumar T, Santhoshkumar S, Sivakumar S (2017) Studies on the occurrence of infectious myonecrosis virus in pond-reared Litopenaeus vannamei (Boone, 1931) in India. J Fish Dis 40(12):1823–1830
Sahul Hameed AS, Anilkumar M, Raj MS, Jayaraman K (1998) Studies on the pathogenicity of systemic ectodermal and mesodermal baculovirus and its detection in shrimp by immunological methods. Aquaculture 160(1-2):31–45
Saleh M, El-Matbouli M (2015) Rapid detection of Cyprinid herpesvirus-3 (CyHV-3) using a gold nanoparticle-based hybridization assay. J Virol Methods 217:50–54. https://doi.org/10.1016/j.jviromet.2015.02.021
Sánchez-Martínez JG, Aguirre-Guzmán G, Mejía-Ruíz H (2007) White spot syndrome virus in cultured shrimp: a review. Aquac Res 38(13):1339–1354
Sánchez-Paz A (2010) White spot syndrome virus: an overview on an emergent concern. Vet Res 41(6):43
Sannino D (2021) Types and Classification of Nanomaterials. In: Tahir MB, Rafique M, Sagir M (eds) Nanotechnology: Trends and Future Applications. Springer Singapore, Singapore, pp 15–38
Sarathi M, Balasubramanian G, Sivakumar VK, Sahul Hameed AS (2008) Artemia is not a vector for monodon baculovirus (MBV) transmission to Penaeus monodon. J Fish Dis 31(8):631–636
Senapin S, Phewsaiya K, Briggs M, Flegel TW (2007) Outbreaks of infectious myonecrosis virus (IMNV) in Indonesia confirmed by genome sequencing and use of an alternative RT-PCR detection method. Aquaculture 266(1-4):32–38
Shi CY, Wang YG, Yang SL, Huang J, Wang QY (2004) The first report of an iridovirus-like agent infection in farmed turbot, Scophthalmus maximus, in China. Aquaculture 236(1-4):11–25
Shike H, Dhar AK, Burns JC, Shimizu C, Jousset FX, Klimpel KR, Bergoin M (2000) Infectious hypodermal and hematopoietic necrosis virus of shrimp is related to mosquito brevidensoviruses. Virology 277(1):167–177
Shiu JY, Hong JR, Ku CC, Wen CM (2018) Complete genome sequence and phylogenetic analysis of megalocytivirus RSIV-Ku: A natural recombination infectious spleen and kidney necrosis virus. Arch Virol 163(4):1037–1042
Skliris GP, Richards RH (1999) Nodavirus isolated from experimentally infected tilapia, Oreochromis mossambicus (Peters). J Fish Dis 22(4):315–318
Somamoto T, Nakanishi T (2020) Mucosal delivery of fish vaccines: Local and systemic immunity following mucosal immunisations. Fish Shellfish Immunol 99:199–207. https://doi.org/10.1016/j.fsi.2020.01.005
Song JY, Kitamura SI, Jung SJ, Miyadai T, Tanaka S, Fukuda Y, Kim SR, Oh MJ (2008) Genetic variation and geographic distribution of megalocytiviruses. J Microbiol 46(1):29–33

Stentiford GD, Bonami JR, Alday-Sanz V (2009) A critical review of susceptibility of crustaceans to Taura syndrome, Yellowhead disease and White Spot Disease and implications of inclusion of these diseases in European legislation. Aquaculture 291(1-2):1–17

Stentiford GD, Neil DM, Peeler EJ, Shields JD, Small HJ, Flegel TW, Vlak JM, Jones B, Morado F, Moss S, Lotz J (2012) Disease will limit future food supply from the global crustacean fishery and aquaculture sectors. J Invertebr Pathol 110(2):141–157

Subramaniam K, Shariff M, Omar AR, Hair-Bejo M (2012) Megalocytivirus infection in fish. Rev Aquac 4(4):221–233

Sudhakaran R, Ishaq Ahmed VP, Haribabu P, Mukherjee SC, Sri Widada J, Bonami JR, Sahul Hameed AS (2007) Experimental vertical transmission of Macrobrachium rosenbergii nodavirus (MrNV) and extra small virus (XSV) from brooders to progeny in Macrobrachium rosenbergii and Artemia. J Fish Dis 30(1):27–35

Sudhakaran R, Musthaq SS, Haribabu P, Mukherjee SC, Gopal C, Hameed AS (2006a) Experimental transmission of Macrobrachium rosenbergii nodavirus (MrNV) and extra small virus (XSV) in three species of marine shrimp (Penaeus indicus, Penaeus japonicus and Penaeus monodon). Aquaculture 257(1-4):136–141

Sudhakaran R, Yoganandhan K, Ahmed VI, Hameed AS (2006b) Artemia as a possible vector for Macrobrachium rosenbergii nodavirus (MrNV) and extra small virus transmission (XSV) to Macrobrachium rosenbergii post-larvae. Dis Aquat Org 70(1-2):161–166

Suganthi G, UMA A, REBECCA G, SARAVANABAVA K (2012) Strain Variations in Monodon Baculovirus (Mbv) Infecting Penaeus Monodon (Tiger Shrimp) in India. CIBTech J Biotechnol 1(1):22–26

Sukenda S, Lila G, Zairin M Jr, Angela L, Alimuddin A (2020) Identification of giant gourami iridovirus (GGIV): a new infectious spleen and kidney necrosis virus (ISKNV) from natural outbreak in cultured Osphronemus goramy. Aquac Int 28(3):1069–1082

Sukhumsirichart W, Attasart P, Boonsaeng V, Panyim S (2006) Complete nucleotide sequence and genomic organization of hepatopancreatic parvovirus (HPV) of Penaeus monodon. Virology 346(2):266–277

Surachetpong W, Janetanakit T, Nonthabenjawan N, Tattiyapong P, Sirikanchana K, Amonsin A (2017) Outbreaks of tilapia lake virus infection, Thailand, 2015–2016. Emerg Infect Dis 23(6):1031

Swaminathan TR, Kumar R, Dharmaratnam A, Basheer VS, Sood N, Pradhan PK, Sanil NK, Vijayagopal P, Jena JK (2016) Emergence of carp edema virus in cultured ornamental koi carp, Cyprinus carpio koi, in India. J Gen Virol 97(12):3392–3399

Taengphu S, Sangsuriya P, Phiwsaiya K, Debnath PP, Delamare-Deboutteville J, Mohan CV, Dong HT, Senapin S (2020) Genetic diversity of tilapia lake virus genome segment 1 from 2011 to 2019 and a newly validated semi-nested RT-PCR method. Aquaculture 526:735423

Takahashi Y, Itami T, Kondo M, Maeda M, Fujii R, Tomonaga S, Supamattaya K, Boonyaratpalin S (1994) Electron microscopic evidence of bacilliform virus infection in kuruma shrimp (Penaeus japonicus). Fish Pathol 29(2):121–125

Tang KF, Durand SV, White BL, Redman RM, Mohney LL, Lightner DV (2003) Induced resistance to white spot syndrome virus infection in Penaeus stylirostris through pre-infection with infectious hypodermal and hematopoietic necrosis virus—a preliminary study. Aquaculture 216(1-4):19–29

Tang KF, Pantoja CR, Lightner DV (2008) Nucleotide sequence of a Madagascar hepatopancreatic parvovirus (HPV) and comparison of genetic variation among geographic isolates. Dis Aquat Org 80(2):105–112

Tang KF, Pantoja CR, Poulos BT, Redman RM, Lightner DV (2005) In situ hybridization demonstrates that Litopenaeus vannamei, L. stylirostris and Penaeus monodon are susceptible to experimental infection with infectious myonecrosis virus (IMNV). Dis Aquat Org 63(2-3):261–265

Tang KFJ, Bondad-Reantaso MG, Surachetpong W, Dong HT, Fejzic N, Wang Q, Wajsbrot N, Hao B (2021) Tilapia lake virus disease strategy manual, vol 1220. Food & Agriculture Organisation

Tattiyapong P, Dachavichitlead W, Surachetpong W (2017) Experimental infection of Tilapia Lake Virus (TiLV) in Nile tilapia (Oreochromis niloticus) and red tilapia (Oreochromis spp.). Vet Microbiol 207:170–177

Thangaraj RS, Nithianantham SR, Dharmaratnam A, Kumar R, Pradhan PK, Thangalazhy Gopakumar S, Sood N (2021) Cyprinid herpesvirus-2 (CyHV-2): a comprehensive review. Rev Aquac 13(2):796–821

Thilsted SH, Thorne-Lyman A, Webb P, Bogard JR, Subasinghe R, Phillips MJ, Allison EH (2016) Sustaining healthy diets: The role of capture fisheries and aquaculture for improving nutrition in the post-2015 era. Food Policy 61:126–131

Thirumalaikumar E et al (2021) Oral delivery of pVAX-OMP and pVAX-hly DNA vaccine using chitosan-tripolyphosphate (Cs-TPP) nanoparticles in Rohu, (Labeo rohita) for protection against Aeromonas hydrophila infection. Fish Shellfish Immunol 115:189–197. https://doi.org/10.1016/j.fsi.2021.06.004

Thwaite R et al (2018) Protein Nanoparticles Made of Recombinant Viral Antigens: A Promising Biomaterial for Oral Delivery of Fish Prophylactics. Front Immunol 9:1652. https://doi.org/10.3389/fimmu.2018.01652

Umesha RK, Uma A, Otta SK, Karunasagar I, Karunasagar I (2003) Detection by PCR of hepatopancreatic parvovirus (HPV) and other viruses in hatchery-reared Penaeus monodon postlarvae. Dis Aquat Org 57(1-2):141–146

Van Hulten MC, Reijns M, Vermeesch AM, Zandbergen F, Vlak JM (2002) Identification of VP19 and VP15 of white spot syndrome virus (WSSV) and glycosylation status of the WSSV major structural proteins. J Gen Virol 83(1):257–265

Van Hulten MC, Witteveldt J, Peters S, Kloosterboer N, Tarchini R, Fiers M, Sandbrink H, Lankhorst RK, Vlak JM (2001a) The white spot syndrome virus DNA genome sequence. Virology 286(1):7–22

Van Hulten MC, Witteveldt J, Snippe M, Vlak JM (2001b) White spot syndrome virus envelope protein VP28 is involved in the systemic infection of shrimp. Virology 285(2):228–233

Venkatesan C, Vimal S, Hameed ASS (2013) Synthesis and Characterization of Chitosan Tripolyphosphate Nanoparticles and its Encapsulation Efficiency Containing Russell's Viper Snake Venom. J Biochem Mol Toxicol 27(8):406–411. https://doi.org/10.1002/jbt.21502

Verma DK, Sood N, Paria A, Swaminathan TR, Mohan CV, Rajendran KV, Pradhan PK (2022) Reassortment and evolutionary dynamics of tilapia lake virus genomic segments. Virus Res 308:198625

Vijayan, K.K., Alavandi, S.V., Rajendran, K.V. and Alagarswami, K., 1995. Prevalence and Histopathology of Monodon Baculovirus (MBV) infection in Penaeus monodon and P. indicus in shrimp farms in the south–east coast of India.

Vimal S et al (2014) Delivery of DNA vaccine using chitosan–tripolyphosphate (CS/TPP) nanoparticles in Asian sea bass, Lates calcarifer (Bloch, 1790) for protection against nodavirus infection. Aquaculture 420-421:240–246. https://doi.org/10.1016/j.aquaculture.2013.11.017

Walker PJ, Winton JR (2010) Emerging viral diseases of fish and shrimp. Vet Res 41(6):51

Waltzek TB, Kurobe T, Goodwin AE, Hedrick RP (2009) Development of a polymerase chain reaction assay to detect cyprinid herpesvirus 2 in goldfish. J Aquat Anim Health 21(1):60–67

Wang Q, Ji W, Xu Z (2020) Current use and development of fish vaccines in China. Fish Shellfish Immunol 96:223–234. https://doi.org/10.1016/j.fsi.2019.12.010

Wang, C.S., Chang, J.S., Shih, H.H. and Chen, S.N., 2006. RT-PCR amplification and sequence analysis of Macrobrachium rosenbergii nodavirus and extra small virus (XSV) associated with white tail disease of M. rosenbergii (de Man) cultured in Taiwan. GenBank Direct Submission

Wang Y, Jehle JA (2009) Nudiviruses and other large, double-stranded circular DNA viruses of invertebrates: new insights on an old topic. J Invertebr Pathol 101(3):187–193

Way K, Stone D (2013) Emergence of carp edema virus-like (CEV-like) disease in the UK. CEFAS Finfish News 15:32–34

Widada JS, Bonami JR (2004) Characteristics of the monocistronic genome of extra small virus, a virus-like particle associated with Macrobrachium rosenbergii nodavirus: possible candidate for a new species of satellite virus. J Gen Virol 85(3):643–646

Xu J, Zeng L, Zhang H, Zhou Y, Ma J, Fan Y (2013) Cyprinid herpesvirus 2 infection emerged in cultured gibel carp, Carassius auratus gibelio in China. Vet Microbiol 166(1-2):138–144

Yang YT, Lee DY, Wang Y, Hu JM, Li WH, Leu JH, Chang GD, Ke HM, Kang ST, Lin SS, Kou GH (2014) The genome and occlusion bodies of marine Penaeus monodon nudivirus (PmNV, also known as MBV and PemoNPV) suggest that it should be assigned to a new nudivirus genus that is distinct from the terrestrial nudiviruses. BMC Genomics 15(1):1–24

Yoganandhan K, Leartvibhas M, Sriwongpuk S, Limsuwan C (2006) White tail disease of the giant freshwater prawn Macrobrachium rosenbergii in Thailand. Dis Aquat Org 69(2-3):255–258

Yoshikoshi K, Inoue K (1990) Viral nervous necrosis in hatchery-reared larvae and juveniles of Japanese parrotfish, Oplegnathus fasciatus (Temminck & Schlegel). J Fish Dis 13(1):69–77

Zhang C et al (2020) Application of Biomimetic Cell-Derived Nanoparticles with Mannose Modification as a Novel Vaccine Delivery Platform against Teleost Fish Viral Disease. ACS Biomat Sci Eng 6(12):6770–6777. https://doi.org/10.1021/acsbiomaterials.0c01302

Zhang H, Zeng L, Fan Y, Zhou Y, Xu J, Ma J (2014) A Loop-Mediated Isothermal Amplification Assay for Rapid Detection of Cyprinid Herpesvirus 2 in Gibel Carp (Carassius auratus gibelio). Sci World J 2014:716413. https://doi.org/10.1155/2014/716413

Zhang W et al (2021) pH-Controlled Release of Antigens Using Mesoporous Silica Nanoparticles Delivery System for Developing a Fish Oral Vaccine. Front Immunol 12:644396. https://doi.org/10.3389/fimmu.2021.644396

Zhao Z et al (2020) Single-walled carbon nanotubes as delivery vehicles enhance the immunoprotective effect of an immersion DNA vaccine against infectious spleen and kidney necrosis virus in mandarin fish. Fish Shellfish Immunol 97:432–439. https://doi.org/10.1016/j.fsi.2019.12.072

Zhao R, Geng Y, Yu Z, Wang K, OuYang P, Chen D, Huang X, He C, Peng G, Lai W (2019) New detection of Cyprinid herpesvirus 2 associated with mass mortality in colour crucian carp (Carassius auratus), in China. Aquac Res 50(6):1705–1709

Zhu B, Zhang C, Zhao Z, Wang GX (2020) Targeted Delivery of Mannosylated Nanoparticles Improve Prophylactic Efficacy of Immersion Vaccine against Fish Viral Disease. Vaccine 8(1):87. https://doi.org/10.3390/vaccines8010087

# Index




V. K. Arivarasan et al. (eds.), *Nanotechnological Approaches to the Advancement of Innovations in Aquaculture*, Nanotechnology in the Life Sciences, https://doi.org/10.1007/978-3-031-15519-2

MIX
Papier aus verantwortungsvollen Quellen
Paper from responsible sources
FSC® C105338

If you have any concerns about our products,
you can contact us on
**ProductSafety@springernature.com**

In case Publisher is established outside the EU,
the EU authorized representative is:
**Springer Nature Customer Service Center GmbH**
**Europaplatz 3, 69115 Heidelberg, Germany**

Printed by Libri Plureos GmbH
in Hamburg, Germany